Marine Algal Bloom: Characteristics, Causes and Climate Change Impacts

Santosh Kumar Sarkar

Marine Algal Bloom: Characteristics, Causes and Climate Change Impacts

Santosh Kumar Sarkar
Department of Marine Science
University of Calcutta
Calcutta, India

ISBN 978-981-13-4103-8 ISBN 978-981-10-8261-0 (eBook)
https://doi.org/10.1007/978-981-10-8261-0

Softcover re-print of the Hardcover 1st edition 2018

Printed on acid-free paper

This Springer imprint is published by the registered company Springer Nature Singapore Pte Ltd.
The registered company address is: 152 Beach Road, #21-01/04 Gateway East, Singapore 189721, Singapore

Dedicated to my parents

Preface

Algal blooms, a recurrent phenomenon in freshwater and marine environment, have been of major global interest as this biological nuisance brings a tremendous negative impact on the primary productivity and sustenance of other pelagic organisms. The occurrence of bloom is triggered by eutrophication, i.e., enrichment with chemical nutrients, typically total nitrogen (TN) and total phosphorous (TP). There is utmost need for better understanding of the effects of algal blooms on seafood quality, the complex physicochemical and biological interactions and subsequent trophodynamics in order to develop the strategy for effective coastal zone management. The book will provide a wealth of exhaustive information encompassing wide spectrum of important and upgraded topics relevant to algal blooms in regional and global scale. The book endeavors to exemplify the characteristics and causes of algal blooms, presenting very basic concepts of this phenomenon (such as Redfield ratio, eutrophication, hypoxia, upwelling, and downwelling) as well as illustrative and advanced account of harmful algal blooms (HABs) followed by the regional case studies from vulnerable coastal as well as mangrove regions showing the negative impact on plankton community structure and finally impact of climate change on the occurrence of algal bloom. Each chapter has its own identity, relevance, and importance and is beneficial for wide audience to get access to advanced knowledge of algal blooms in marine environments as a whole.

An exclusive chapter on "Harmful Algal Blooms (HABs)" (commonly known as "red tides") has been devoted which is an emerging and potential natural ecological and economic disasters, posing serious risks to human health, environmental sustainability, and aquatic life (including endangered marine mammals). HABs refer to a rapid proliferation of phytoplankton species such as dinoflagellates, diatoms, and cyanobacteria in aquatic ecosystems. Nowadays, there are about 90 marine planktonic microalgae capable of producing toxins and most of them are dinoflagellates, followed by diatoms.

Considering its sheer negative impact of HABs in a regional and global scale, I have given illustrative account of the following important topics: dinoflagellates – an ideal potential agent for HAB; red tide and HABs; and five diverse categories, such as paralytic shellfish poisoning (PSP), neurotoxic shellfish poisoning (NSP), amnesic shellfish poisoning (ASP), diarrhetic shellfish poisoning (DSP), and ciguatera fish poisoning (CFP) (seafood poisoning). HABs are seriously harmful because of the vast array of biotoxins or "environmental chemicals" that they can produce which can poison humans, land and aquatic mammals, fish, shellfish, and other aquatic species. Most significantly, these toxins interfere with human and animal metabolism, nerve conduction, and central nervous system processing of information, as illustrated in subsequent chapters. The information would certainly add value for better understanding of the occurrence and impact of HABs, which seem to be very complex and of global interest.

The unique coupling of the three elements, eutrophication–HABs– biodiversity, is challenging paradigms in a world of complex nutrient change. The global HAB problem is on a trajectory for more blooms, more toxins and more frequent in more coastal regions. In general, climate change is expected to result by the key environmental factors such as higher sea water temperatures, light availability, nutrients, more stratification, and hence more potentially harmful dinoflagellates and cyanobacteria. Other climate change effects such as wind and precipitation patterns, altered oceanic circulation patterns and carbon dioxide concentrations further complicate regional effects of climate change.

The book is intended to serve as a useful, reliable, and upgraded reference source for a large section of people, such as coastal managers, students and researchers, policymakers, and environmentalists, involving in coastal research and management at regional and global scale. I trust that the book would provide better understanding and stimulate greater interest in the topic metal pollution and educated public would come forward in taking dynamic role to reduce pollution as a whole. Constructive comments and suggestions for improvement of the text are gratefully appreciated.

Keywords Harmful algal blooms · HABs · Red tide · Management · Control · Monitoring

Calcutta, India Santosh Kumar Sarkar

Acknowledgments

I acknowledge the generous support from the following academicians and researchers who have shared their expertise and provided me with their constructive suggestions and valuable comments: Drs. Jiang-Shiou Hwang, National Taiwan Ocean University; Simonetta Corsolini, University of Siena, Italy; Krishna Das, University of Liege, Belgium; Karla Pozo, Masaryk University, Czech Republic; Emmanoel V. Silva-Filho, Fluminense Federal University, Brazil; Jason Kirby, Commonwealth Scientific and Industrial Research Organization (CSIRO), Australia; M.P. Jonathan, (CIIEMAD) (IPN), Mexico; M.N.V. Prasad, University of Hyderabad, India; Rahul Kundu, Saurashtra University, India; Jayant Kumar Mishra, Pondicherry University, India; Dr. Mathammal Sudarshan, Inter-University Consortium for DAE Facilities, Calcutta Centre, India; and K.K. Satpathy, Indira Gandhi Centre for Atomic Research, (IGCAR), India. The incredible support and technical assistance rendered by my research students, Bhaskar Deb Choudhury, Ranju Choudhury, Dibyendu Rakshit, Soumita Mitra, and Priyanka Mondal, as well as my daughter, Prathama Sarkar, for preparing the manuscript are greatly appreciated.

For contribution of photography, I am greatly indebted to the following scientists:

1. Dr. Mindy L. Richlen, National Office for Harmful Algal Blooms at Woods Hole Oceanographic Institution, USA
2. Dr. Rut Akselman, Instituto Nacional de Investigación y Desarrollo Pesquero, Argentina
3. Dr. Christopher J. Gobler, School of Marine and Atmospheric Sciences, Stony Brook University, New York
4. Dr. Kathryn Taffs, Southern Cross University, School of Environment, Science and Engineering, Australia

I sincerely acknowledge the permission extended by the Editor, *Indian Journal of Geo-Marine Science*, CSIR, New Delhi, India, for partly reproducing the following research papers used in the book: Vol. 32(2), June 2003, pp. 165–167; Vol. 34(2), June 2005, pp. 163–173; Vol. 38(1), March 2009, pp. 77–88; Vol. 39(3),

September 2010, pp. 323–333; Vol. 41(4), August 2012, pp. 304–313; Vol. 43(2), February 2014, pp. 258–262; Vol. 44(9), September 2015, pp. 1282–1293.

Some relevant information was also taken from the following web links:

http://www.ut.ee/_olli/eutr/
https://phys.org/news/2013-07-china-largest-ever-algae-bloom.html
http://www.ibtimes.co.uk
https://products.coastalscience.noaa.gov/pmn/_docs/Factsheets/Factsheet_Azadinium.pdf
http://www.emedicine.com/EMERG/topic528.htm
http://www.algaebase.org
http://www.epd.gov

Finally, I express my sincere gratitude to Dr. Mamta Kapila, Springer (India) Pvt. Ltd., Life Science and Biomedicine, Senior Editor; Dr. Judith Terpos, Senior Editorial Assistant, Environmental Science; Ms. Raman, Daniel Ignatius Jagadisan, Springer Nature, SPi Global, Project Coordinator; Ashok Kumar, Senior Executive and Project Coordinator, Springer Nature, for their keen interest, full support, and constant encouragement in the publication of the book. Last but not the least, I thank my wife, Manjushree Sarkar, who tolerated my incessant work and extended her full support and cooperation in completing the work.

I sincerely hope this would provide greater understanding and stimulates greater interest in the topic metal speciation. Constructive suggestions for improvement of the text are gratefully appreciated and should be addressed to the Springer.

Contents

About the Author

Santosh Kumar Sarkar is Professor of Marine Science, University of Calcutta, India. He has over 30 years of research experience in the field of marine and estuarine ecology, biology, pollution and geochemistry. He has been the Principal Investigator for 12 national projects and 8 international research projects in collaboration with eminent scientists. He has engaged in collaborative research work with leading research institutes around the globe, namely, the Institute Rudjer Boskovic, Croatia; Universities of Genoa Milan, Sienna, and Venice, Italy; IPIMAR/INRB IP, and the University of Trás-os-Montes e Alto Douro, Portugal; University of Edinburgh, British Geological Survey, UK; University of Wuppertal, Germany; CIIEMAD National Politechnic Institute, Mexico; Centre National de la Recherche Scientifique (CNRS), Marine Microbial Ecology, France; Southern Cross University, Australia; Evandro Chagas Institute, Ministry of Health Rodovia BR 316 Km 07 S/N Ananindeua, Pará, Brazil and Vrije Universiteit Brussel, Belgium. He has authored more than 90 research papers in peer-reviewed international journals, edited a book, and contributed several book chapters covering diverse disciplines of marine environment. He has authored the following books: (1) "Loricate Ciliate Tintinnids in a Tropical Mangrove Wetland: Diversity, Distribution and Impact of Climate Change," Springer (2014); (2) "Marine Organic Micropollutants: A Case Study of the Indian Sundarban Mangrove Wetland," Springer (2016); and (3) "Trace Metals in a Tropical Mangrove Wetland: Chemical Speciation, Ecotoxicological Relevance and Remedial Measures" Springer (2017).

Chapter 1
Algal Blooms: Basic Concepts

Abstract An algal bloom is a rapid and prolific increase in phytoplankton biomass in freshwater, brackish water or marine water systems and is recognized by the discolouration in the water based on the phytopigments in the algal cells (either innocuous or toxic). Algae can be considered to be blooming at widely varied concentrations, reaching millions of cells per millilitre, or tens of thousands of cells per litre. Occurrence of bloom and its persistence are a complex environmental process involving multiple factors such as anthropogenic nutrient (eutrophication), solar radiation, temperature, current patterns and other associated factors. This natural but stochastic event leads to severe ecological health hazards, degradation of water quality and productivity and pelagic community structure. Proliferations of toxic microalgae in aquatic systems can cause massive fish kills, contaminate seafood with toxins and alter ecosystems in ways that humans perceive as harmful. The chapter addresses a comprehensive account regarding basic features related to algal bloom, such as Redfield ratios, eutrophication and hypoxia (deoxygenation). This is followed by detailed account of the major bloom causative agents (diatoms, dinoflagellates and cyanobacteria) and the remote sensors usually in practice for water quality monitoring. Finally a comprehensive account of the analytical instruments has been discussed generally used for microalgal studies for taxonomic and chemical component analyses.

Keywords Algal bloom · Eutrophication · Hypoxia · Remote sensors · Water quality · Diatom · Dinoflagellate · Cyanobacteria

1.1 Introduction

The International Council for the Exploration of the Seas (ICES 1984) has defined algal blooms as noticeable changes in water discolouration, foam production and eventually invertebrate mortality or toxicity to humans. It is worth to refer that about 5000 species of marine phytoplankton, ~ 300 species are concerned for algal blooms dominated by diatoms, dinoflagellates and cyanobacteria followed by raphidophytes, prymnesiophytes, cyanophytes, prasinophytes and silicoflagellates. Algal bloom is one of the potential biological nuisances caused by sudden, rapid and

S. K. Sarkar, *Marine Algal Bloom: Characteristics, Causes and Climate Change Impacts*, https://doi.org/10.1007/978-981-10-8261-0_1

perceptible increase in phytoplankton biomass in freshwater, coastal or ocean system. Occurrence and persistence of the bloom are complex environmental issue involving physical, chemical and biological factors. The explosive increase in phytoplankton stock is mainly caused due to enrichment of inorganic nutrients.

However, only a few dozen of these species has the ability to produce potent biotoxins, and widely recognized as 'Harmful Algal Blooms' (HABs), a global alarming phenomenon resulting hypoxia, alternation in community structure affecting the food webs, water quality degradation. Occurrence and geographic extension of HABs, economic disruption and nature of toxins and number of toxic species have also increased significantly (Smayda and White 1990; Hallegraeff 1993; Van Dolah 2000).

In temperate latitudes, a seasonal biological phenomenon in eutrophic (surface layer receiving sufficient sunlight for photosynthesis) coastal waters is exclusively pronounced due to strong increase in phytoplankton biomass (stock) that typically occurs in early spring and lasts till late spring/early summer. A special physical condition (increased penetration of solar radiation in the photic region) is the crucial factor for causing this temporary bloom recorded in the shelf, slope, warm-core rings and Gulf Stream waters. Irradiance exceeds a certain threshold; the production season starts with exponential increase in number of diatoms and flagellates. In contrast, there are no thermal stratifications of the water column during winter. As a result, phytoplankton and nutrients are transported up and down out of the restricted euphotic zone so quickly that phytoplankton gets low levels of light exposure, inhibiting primary productivity level. The surface waters warm up in spring as the sun rises higher and penetrates deeper into ocean. Phytoplankton and nutrients are trapped in this warm surface layers. The euphotic zone also deepens, and the 'spring bloom' evolved due to rapid proliferation of phytoplankton in this favourable environmental conditions. In temperate seas, 50–75% of the annual primary production occurs during the spring blooms in March–April.

1.2 Redfield Ratio or Redfield Stoichiometry

Phytoplankton production demands a variety of both organic and inorganic nutrients, including C, N, P and Si, metals (especially Fe and Cu), and trace elements, along with vitamins. Macronutrients (carbon, nitrogen, phosphorus and silicon) are the most important ones in which N and P are the most important ones among them as they are usually in short supply relative to demand. Hence availabilities of N and P mainly control the rates of phytoplankton-mediated primary production (Ryther and Dunxtan 1971; Nixon 1995, 1996). Again as different algal species require these nutrients in different fixed proportions, their supply ratios and other elements (e.g. N/Si and N/Fe) modulate the growth and competitive interaction among the phytoplankton groups (Syrett 1981; Dortch and Whitledge 1992; Stolte et al. 1994; Justic et al. 1995; Turner et al. 1998).

Redfield (1934) observed a unique consistency in elemental composition of seawater as well as phytoplankton, and this should be treated as most significant contribution in the field of aquatic biogeochemistry. Redfield ratio or Redfield stoichiometry is the characteristic of atomic weights of the elements carbon (C), nitrogen (N) and phosphorus (P). C:N:Si:P are in the following ratio 106:16:15:1, found in phytoplankton and throughout the deep oceans. This balance is not always constant and can be changed by episodic nutrient inputs or depletion from human and natural sources, modulated by rainfall and runoff, groundwater and oceanic inputs. The name of this characteristic ratio is associated with the American oceanographer A.D. Redfield who had analysed thousands of research samples of marine biomass across all of the ocean regions. He ascertained that globally the elemental composition of marine organic matter (both living and dead) was remarkably constant across all of the regions.

The empirically developed stoichiometric ratio of C:N:P remains relatively consistent from both the coastal and open ocean regions, which was originally found to be C:N:P = 106:16:1 and has been recently revised to 117:14:1. The valid reason for this constant molar ratio that occurs in the oceans is not clear. However, the ratio can be altered by nutrient input due to catastrophic events or depletions from a variety of human and natural sources, which are modulated by and delivered to coastal waters by rainfall and runoff, groundwater and oceanic inputs. Murata et al. (2012) demonstrated the cellular toxicities of paralytic shellfish poison (PSP) (described in Chap. 3) in harmful dinoflagellate *A. tamarense* from the N:P ratios, ranging from 4 to 64. The growth and paralytic poison content in this harmful dinoflagellate were also determined by them based on the N:P ratio.

The near coincidence of nitrogen-to-phosphorous ratios in the sea and the phytoplankton requirements has led to the conclusion that growth of phytoplankton cells, followed by their sinking as well as decomposition, controls the N:P ratio in both the phytoplankton and seawater. A balance of nitrogen fixation and denitrification would fix the overall value of nitrogen relative to phosphorous. It is assumed that phosphorous is lost at greater rate from deeper ocean than N which is subject to further studies. However, numerous examples exist of phytoplankton species that deviate from the N:P uptake ratio of 16:1 as referred above.

Certain phytoplankton groups have highly specific nutrient requirements, especially the diatoms, which have siliceous cell walls or frustules. While this provides for strong, morphologically diverse cell walls, it also makes this group. Silicon is a product of weathering of upstream siliceous soils, and its supply may be limited if these soils are absent or present in traces. In addition, the supply ratio of silicon to other potentially limiting nutrients (N and P) may detect the relative availability of one or several nutrients in order to maintain (nutrient) balanced or 'Redfield' growth (Redfield 1958). Hence, if elevated N and/or P loadings occur as a result of human or climatic perturbations, as discussed above, the supply rate of silicon may become limiting, resulting changes in the phytoplankton community structure. These appear to be a case in the northern Gulf of Mexico region, where Mississippi River discharge supplies excess N and P relative to silicon (Justic et al. 1995; Turner et al. 2001).

There is a relationship between diatom and flagellate abundance and N:P:Si ratios, and probability of a flagellate bloom might increase in strongly eutrophicated areas due to silicon limitation of diatom growth. Smayda (1989, 1990), based on an assessment of global coastal waters, demonstrated that eutrophic enrichment of N and P has led to long-term declines in the Si:N and Si:P ratios. He has suggested that the decline in Si:P ratio has particularly favoured non-diatom bloom in response to eutrophication and such ionic ration shifts are a key factor associated with the apparent epidemic of novel blooms and phylogenetic shifts in biomass predominance in the sea.

1.3 Eutrophication

Eutrophication or the anthropogenic elevation of nutrient inputs to an ecosystem (more precisely hypertrophication) is characterized by explosive growth of microalgae (and/or macroalgae) due to the greater availability of one or more plant limiting growth factors necessary for photosynthesis, such as sunlight, carbon dioxide and inorganic nutrients (ammonia, nitrate and phosphate) (Schindler 2006). Eutrophication, in which 'eu' = 'well' or 'good' and 'trope' = 'nourishment', actually implies 'too much of a good thing' which is actually 'bad' as it involves in several large-scale alteration/changes in structure, function and stability of the marine ecosystems. These have direct implications on planktonic microbial food webs in such eutrophicated systems.

This phenomenon also initiates abrupt growth of macroalgae, increased sedimentation and oxygen consumption, oxygen depletion in the bottom water (anoxic/hypoxic condition) and sometimes the death of benthic organisms and fish. This is one of the major stresses on coastal ecosystems, severely deteriorates the water quality and liable to changes dissolved nitrogen and phosphorus ratio in the water, (DIN:DIP ratio), recognized as Redfield ratio (details given in previous section). Significantly lower ratio causes potential nitrogen limitation, while a higher ratio leads to phosphorus limitation of phytoplankton primary production. It can lead to a shift in species composition to fast-growing algae species (including toxic species) and a shift from long-lived macroalgae to more nuisance species. This is also associated with major changes in aquatic community structure (in terms of species diversity, species richness and species evenness), absolute dominance of small-sized zooplankton (e.g. calanoid species *Paracalanus* sp., *Acrocalanus* sp.). The greatest impoverishment of diversity and density of phytoplankton, microzooplankton and mesozooplankton community was recorded during the prolific bloom of the centric diatom *Hemidiscus hardmannianus* (Bacillariophyceae). A bloom of *Noctiluca* sp. was recorded with high density causing red colouration of the surface water in the Sea of Cortez, Mexico (Figs. 1.1.1 and 1.1.2).

Eutrophication is a process driven by the enrichment of water by nutrients, especially compounds of nitrogen and/or phosphorus, which support high biological productivity, leading to absolute negative impact on pelagic as well as benthic systems. This appreciably degrades the function and health of an ecosystem, adversely affecting *losses in biodiversity, ecosystem degradation, harmful algal blooms and oxygen deficiency in bottom waters* and/or the sustainable provision of goods and services. The process also gives birth to the following consequences:

Fig. 1.1.1 A *Noctiluca* bloom in the Sea of Cortez, Mexico. (*Courtesy*: Dr. Mindy L. Richlen, National Office for HABs, WHOI, USA)

1. Reduced penetration of solar radiation received by bottom waters can lead to the reduction in the depth distribution of macroalgae and seagrasses.
2. Increased decomposition of organic matter (dead algae) can lead to oxygen deficiency in bottom waters. Lowered oxygen concentrations can then impact the fish and benthic fauna (animals living on the bottom of the sea or a lake), which either flee or die from the area.
3. Cause a shift in the biodiversity and ecosystem balance.
4. Cause shellfish poisoning in humans and be of danger to livestocks in coastal water.
5. Water quality is degraded due to decaying algae with foul odours and foam on beaches, or toxins from blooms, impacting the tourism industry.

Scientists linked algal blooms to nutrient enrichment resulting from anthropogenic activities such as agriculture, industry and sewage disposal (Schindler 1974).

Estuarine waters may become hypoxic (oxygen poor; <3 ml per litre) or anoxic (complete lack of oxygen) from algal blooms. While hypoxia may cause animals in estuaries to become physically stressed, anoxic conditions can drastically reduce fish and shellfish stocks and thereby have an economic impact on the fishery industry.

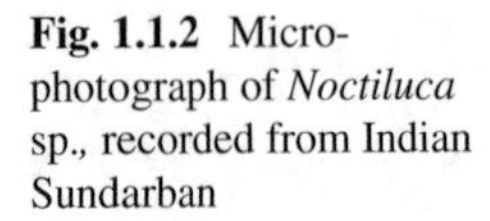

Fig. 1.1.2 Microphotograph of *Noctiluca* sp., recorded from Indian Sundarban

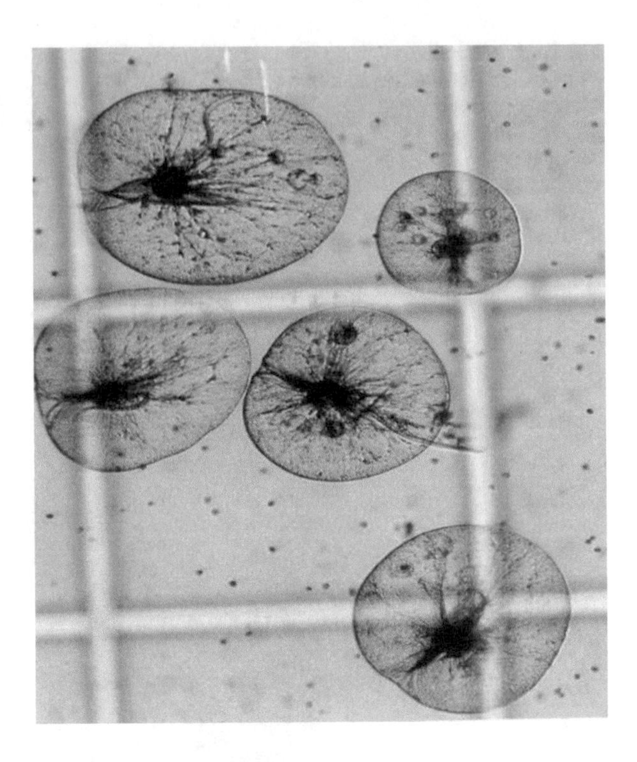

The impacts of eutrophication include the generation of oxygen-depleted or 'dead zones', increased occurrence of microorganisms pathogenic to marine life and even humans and the promotion, usually in combination with other human stressors, of plagues of gelatinous zooplankton. HABs are also commonly recognized as red tides and brown tides (discussed in details in Chap. 3) that can release very powerful toxins into the water. Toxic algal blooms disrupt tourism due to foul and unhygienic odours, and poisoned fish and shellfish adversely affect recreational and commercial fisheries (Carpenter et al. 1998; Howarth et al. 2000) with socio-economic impacts.

Nitrogen and phosphorous are the primary inorganic nutrients responsible for the eutrophication of marine waters, and both of them occur naturally in marine waters, transferred from land via streams, rivers and runoff of rainwater and also from degradation of organic material within the water. However, human inputs of nutrients to the environment have increased the load of nitrogen and phosphorous to the oceans. Most excess nutrients come from discharges of sewage treatment plants and septic tanks, storm water runoff, discharge from sewage treatment plans, municipal and industrial waste waters, agricultural and urban runoff and dredging operations.

The predominant nitrogen load originates from diffuse sources on land, especially agricultural areas along with aquaculture, waste water treatment plants, industrial water and adjacent oceans. Nitrogen is primary responsible for the eutrophication of temperate estuaries and coastal regions, whereas phosphorus is responsible for many tropical estuarine and coastal systems.

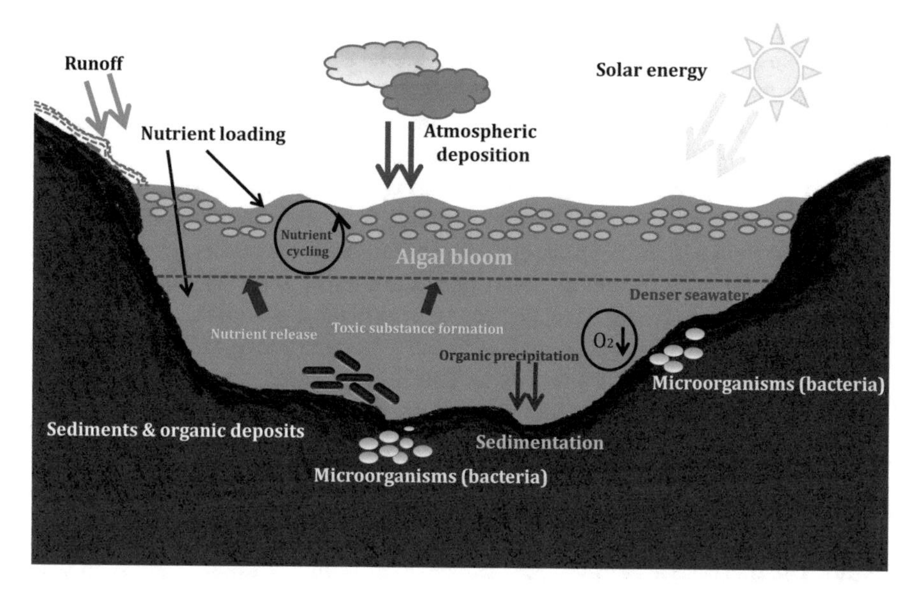

Fig. 1.2 Schematic representation of the mechanism for eutrophication condition in marine environment

The main contribution of human-introduced phosphate comes from domestic and industrial sewage and waste water. Runoff from land is also an important source of phosphate as it is a component of fertilizers. The formation mechanism of eutrophication process has been depicted in Fig. 1.2.

When algae start to grow in an uncontrolled manner, an increasingly large biomass is formed which is destined to degrade. In deep water, a large amount of organic substance accumulates, represented by the algae having reached the end of their life cycle. To destroy all dead algae, an excessive consumption of oxygen is required, in some cases almost total, by microorganisms. An anoxic (oxygen-free) environment is thus created on the lake bottom, with the growth of organisms capable of living in the absence of oxygen (anaerobic), responsible for the degradation of the biomass. The microorganisms, decomposing the organic substance in the absence of oxygen, free compounds that are toxic, such as ammonia and hydrogen sulphide (H_2S). The absence of oxygen reduces biodiversity causing, in certain cases, even the death of animal and plant species. All this happens when the rate of degradation of the algae by microorganisms is greater than that of oxygen regeneration, which in summer is already present in low concentrations.

Water quality parameters (inorganic and organic nutrient concentrations, oxygen and chlorophyll availability and biological oxygen demand) are not the exclusive indicator for marine eutrophication, but benthic status and process parameters are also responsible, such as relative cover and growth rates of indicator algae, invertebrate recruitment, sedimentary oxygen demand and interactions between indicator organisms. The primary future challenge lies in understanding the interaction between marine eutrophication and the two main marine consequences of climate change, ocean warming and acidification. Management action should focus on increasing the efficiency of nutrient usage in industry and agriculture while at the same time minimizing the input of nutrients into marine ecosystems in order to mitigate the negative effects of eutrophication on the marine realm.

Due to decomposition of huge biomass during eutrophication event, depletion of oxygen and mass mortality are generally found which leads to an increased risk of complete system deterioration. In majority of the cases, the prolific phytoplankton growth leads to shifts in species diversity of the algal community, which becomes dominant for a short time period as it absorbs nutrients most quickly. Surprisingly, eutrophication caused by enrichment of N and P leads to a shift in the system from diatoms to dinoflagellates, eventually making a pronounced change in food chain. This is unfavourable for sustenance of other existing species.

Increased input of nutrients, mainly nitrogen (N) and phosphorus (P), to marine waters – with eutrophication as consequence – is a worldwide phenomenon (Cloern 2001). The supply of these nutrients is seldom in optimal balanced ratios (N:P = 16, Redfield ratio), thus resulting in nutrient-limited growth of phytoplankton (Conley 1999). The most obvious effect of eutrophication is an alteration of the food web that results in high levels of phytoplankton biomass, which can lead to algal blooms (Anderson et al. 2002; Smayda 2004). Such shifts in the primary producer community are likely to cascade through the pelagic food web. Disruption of the food web occurs by selection for opportunistic, inedible phytoplankton species, with a poor nutritional value or unsuitable size in relation to the optimal diet of zooplankton. Phytoplankton succession, however, cannot be explained solely by abiotic factors, acting together with competition and grazing; chemically mediated interactions should also be considered (Lucas 1947).

Eutrophication is one of the key local stressors for coastal marine ecosystems and is a major global challenge to the scientists and policymakers because the physical and biological processes linking nutrients and their impacts are complex. The huge influx of land-derived inorganic nutrients (e.g. nitrate and phosphate), along with particulate and dissolved organic matter (POM and DOM), affects the growth and physiology of marine organisms. A growing number of estuarine and coastal marine regions show signs of eutrophication with negative impact on ecosystem functioning. Hence systematic and careful monitoring is required, and nutrient management strategies must be formulated to develop effective and ecofriendly solutions to eutrophication.

1.3.1 Dinoflagellate Cyst as Biosignal for Eutrophication: A Case Study

Dinoflagellate cysts, which are hypnozygotes produced during their sexual life cycle, play a crucial role to protect the microalgae from stressed environment. Cyst germination is considered as a reliable source of causes of HABs and provides alternate source of information for predictable harmful algal blooms in marine environments. These dinoflagellate cysts have been used for understanding palaeoenvironmental changes, especially eutrophication process caused by natural and anthropogenic stresses.

Two major signals for eutrophication are predictable using the dinoflagellate cysts: (1) the Oslofjord signal which is characterized by a remarkable increase in total cyst densities accompanied by increases in a single species such as the autotrophic *Lingulodinium machaerophorum* in the case of Oslofjord, Norway, and (2) the Heterotroph signal which is indicated by dominance of heterotrophic species such as the cysts of *Polykrikos kofoidii/schwartzii*, cysts of *Protoperidinium* spp. and/or cysts of the diplopsalid in Tokyo Bay and Apponagansett Bay in Massachusetts, USA, as these dinoflagellates can consume autotrophic and heterotrophic microplanktonic organisms (Dale 2009). The relationship between eutrophication and increases of both autotrophic and heterotrophic dinoflagellate cysts involving both the above-referred signals can be explained. However, in order to employ these two signals for understanding other geographical regions, more adequate and relevant data such as species-specific relationships and their correlations with nutrient limitation levels are also required.

Ismael et al. (2014) provided detailed and valuable information about the dinoflagellate cyst assemblages as an indication of eutrophication in the Eastern Harbour of Alexandria, Egypt, which has a well-documented history of cultural eutrophication from sewage discharge. Dinoflagellate cysts are an important group of microfossils with a potential as biological indicators of the timing and degree of environmental changes in estuaries (Pospelova and Head 2002). Approximately 13–16% of living dinoflagellates and 10% of harmful dinoflagellates produce resting cysts during their life cycle (Head 1996; Dale 1983). Organic nature of the walls of many cysts makes them resistant to degradation (Fensome et al. 1993), providing a good fossil record. Bottom sediments accumulating in aquatic systems represent an archive of past environmental change, and both geochemical and micropaleontological methods are being developed in the search for eutrophication signals (Dale 2009). Assemblages of dinoflagellate cysts accumulated in sediments can encode information about dinoflagellate population in the upper water column (Dale 1976). Thus, the assemblage of dinoflagellate cysts in the sediment reflects the ecology of living dinoflagellates, the latter influenced by environmental factors such as temperature, salinity, nutrients, diatom availability and turbidity (Taylor and Pallingher 1987). Marret and Zonneveld (2003) summarized the global distribution of extent organic-walled dinoflagellate cysts showing the relationship between distribution pattern of cyst species and surface water parameters. Approximately half of dinoflagellates are heterotrophic, and half are autotrophic (Marret and Zonneveld 2003; Dale 1996). Earlier works on the phytoplankton of the Egyptian Mediterranean waters were all based on the study of the vegetative cells. A little is known about the distribution and occurrence of dinoflagellate cysts. Although the work of Ismael and Khadr (2003) was the first of the dinoflagellate cysts in the E.H, they studied only the distribution of *Alexandrium minutum* cysts in the upper 18 cm of the harbour sediment. The work has highlighted to use the assemblages of dinoflagellate cysts in core sediments as indicator of eutrophication in the Eastern Harbour of Alexandria which has a well-documented history of cultural eutrophication from sewage discharge.

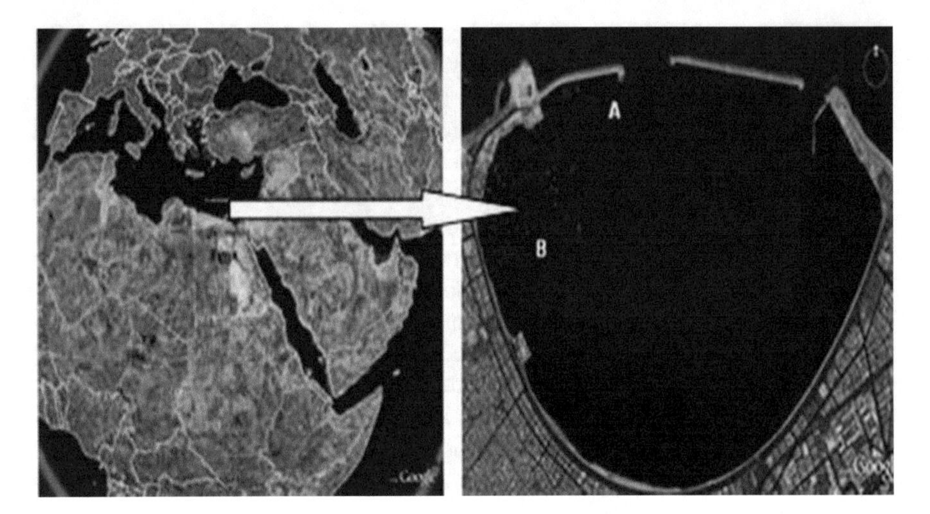

Fig. 1.3 Sampling site in the Eastern Harbour of Alexandria, Egypt

1.3.1.1 Methodology

Two sediment core samples of 78 cm and 56 cm, respectively, were collected from the Eastern Harbour during June 2008 (Fig. 1.3) with the help of scuba divers. Cores were immediately refrigerated, divided into two halves, each half sliced into 2 cm interval and then preserved in labelled sterilized plastic bags until cyst analysis is done. Core sediments were subjected to grain size analysis (Folk and Ward 1957), and the pipette analysis was used for the fine fractions (more than 4Φ) using the technique described by Krumbein and Pettijohn (1938). The organic carbon content was determined according to EL Wakeel and Riley (1957), whereas total phosphorous was determined according to Aspila et al. (1976) method and TN determined using Kjeldahl method (Mudroch et al. 1997).

For cyst analysis two different methods were used for cleaning and concentrating cysts from sediments. In the first method, sediment was processed palynologically (Matsuoka and Fukuyo 2000). Each 2 cm interval of sediments was treated with 10% hydrochloric acid to remove shell fragments. After a series of washing, sediments were passed through 125 mm then 20 mm mesh sieves, and residues from the 20 mm mesh sieves were collected. In the other method, sieving technique was used where 2 cm slice of core sediment was suspended into filtered seawater. Ultrasonic bath was used to disaggregate sediments and cysts. Larger particles were removed by sieving through 250 μm and 125 μm sieve, and the cysts were collected on to a 20 μm sieve (Matsuoka and Fukuyo 2000). One mL duplicate samples were withdrawn for cyst counting using Sedgewick-Rafter chamber and analysed under a bright-field microscope from 100× to 400× magnification, final unit being cysts per gram wet sediment (Cyst g^{-1} wet wt). Cysts were identified to species level using paleontological taxonomy as adapted from Matsuoka (Matsuoka 1985, 1987). In addition, diatoms were counted using Zeiss inverted microscope. For statistical analysis, correlation coefficient was calculated between the pairs of measured variables using programme Statistica 0.7.

1.3.1.2 Cyst Population

Dinocyst preservation in marine sediment is generally adequate for observation, and identification at species level even in the lowest samples identified cysts belonged to six groups, namely, Gonyaulacoid, Protoperidinium, Tuberculodinioid, Calciodinelid, Diplopsalid and Gymnodinioid. Fifteen different genera representing twenty-six species of dinoflagellate cysts were recognized and counted in the palynological assemblages from the Eastern Harbour cores (Table 1.1). Of the

Table 1.1 List of dinoflagellate cyst species recorded from the core sediments in the Eastern Harbour of Alexandria, Egypt

Cyst species (paleontological name)	Motile species (biological name)
Goniaulacoid group	Order Gonyaulacles
Alexandrium sp. 1 (ellipsoidal)	*Alexandrium tamarense* (Lebour) Balech
	A. catanella (Whedon and Kofoid) Balech
Alexandrium sp. 2 (ovoid)	*A.minutum* Halim
Linglodinium machraerophorum	*Lingulodinium polyedra* (Stein) Dodge
Spiniferites bulloideus	*Gonyaulax scrippsae* Kofoid
Spiniferites ramosus	*G. spinifera* (Claparede and Lachmann) Diesing
Spiniferites sp.	*G. verior* Sournia
Operculodinium centrocarpum	*Protoceratium reticulatum* (Claparede and Lachmann) Bütschli
Tuberculodinoid group	Family Pyrophaceae
Tuberculodiniumvancampoae	*Pyrophacus horologium* Stein
Protoperidinoid group	Order Periniales
Selenopemphix alticintum	*Protoperidinium subinerme* (Paulsen) Loeblich II
S. quanta	*P. conicum* (Gran) Balech
Votadinium calvum	*P. oblongum* (Aurivillius) Parke and Dodge
Brigantedinium minutum	*P. minutum* (Kofoid) Loeblich II
Brigantedinium spp.	*Protoperidinium* sp. 1
Brigantedinium spp.	*Protoperidinium* sp. 2
Quinquecuspis sp.	*Peridinium quinquecorne* Abe
Diplopsalid group	Family Diplopalidaceae
Diplopelta parva	*Diplopelta parva* (Abe) Matsuoka
Diplopelta sp.	*Diplopelta* sp.1
Diplopsalislenticula	*Diplopsalislenticula* Bergh
Diplopsalis sp.	*Preperidinium lenticulatum* (Pavillard) Elbrächter
Calciodinelid group	Family Calciodinellaceae
Scrippsiella trochoidea	*Scrippsiella trochoidea* (Stein) Loeblich III
S. *Crystalline*	*S. crystallina* Lewis
Gymnodinoid group	Order Gymnodiniales
Gymnodinium sp.	*Gymnodinium catenatum* Graham
Gymnodinium sp.	*Akashiwo sanguinea* (Hirasaka) Hansen et Moestrup
Gyrodinium sp.	*Gyrodiniumin striatum* Freudenthal and Lee
Polykrikos reticulatum	Family Polykrikaceae
	Polykrikos schwartzii Bütschli

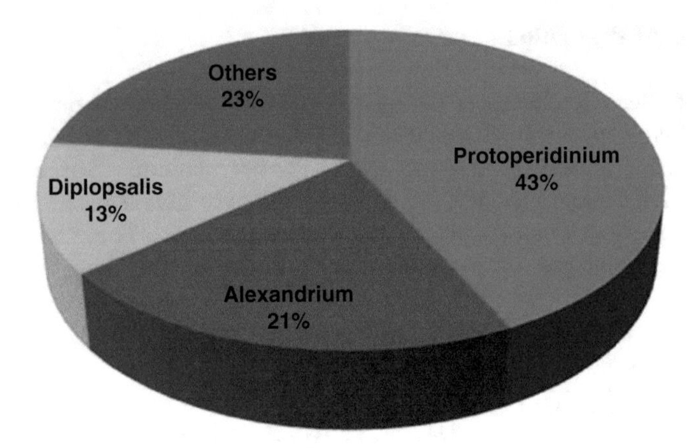

Fig. 1.4 Percentage composition of dinoflagellate cysts from core sediments in the Eastern Harbour of Alexandria, Egypt

individual cyst types, *Protoperidinium* was the most abundant at the two cores, contributing a maximum of 43%, followed by *Alexandrium* 21% and *Diplopsalis* with 13% (Fig. 1.4). Heterotrophic dinoflagellate cysts represent 38% of the total number of dinoflagellate cysts.

Five cyst types have been recorded for the first time in the E.H. Although they were found in deep sediment cores, they are not recorded from the phytoplankton samples in previous studies, *Protoceratium reticulatum*, *Polykrikos schwartzii*, *Alexandrium tamarense*, *A. affine* and *Scrippsiella crystalline*, in addition to one sand-dwelling dinoflagellate and one benthic dinoflagellate: *Sinophysis* sp. and *Prorocentrum lima*.

1.3.1.3 Cyst Distribution

In core A, dinoflagellate cyst concentrations are mostly constant with no remarkable change throughout the core. Peaks recorded did not exceed 574 cyst [g^{-1}] [s4], except in the upper layer (0–2 cm), where it reached its maximum (859 cyst [g^{-1}] [s5]). While in core B, the total cyst concentrations showed two layers. Upper layer from 0–16 cm ranged from 240 to 1296 cyst [g^{-1}] [s6], the lower layer (16–56 cm) being much poorer ranging from 0 to 534 cyst [g^{-1}] [s7] (Fig. 1.5).

Although the abundance of cysts is not necessarily proportional to the magnitude of the respective blooms because of the high variability and uncertainty in the sedimentation rate, dinoflagellate cyst assemblages recorded from the core sediment correspond to the vegetative blooms in the water column as reported in previous works (Table 1.2).

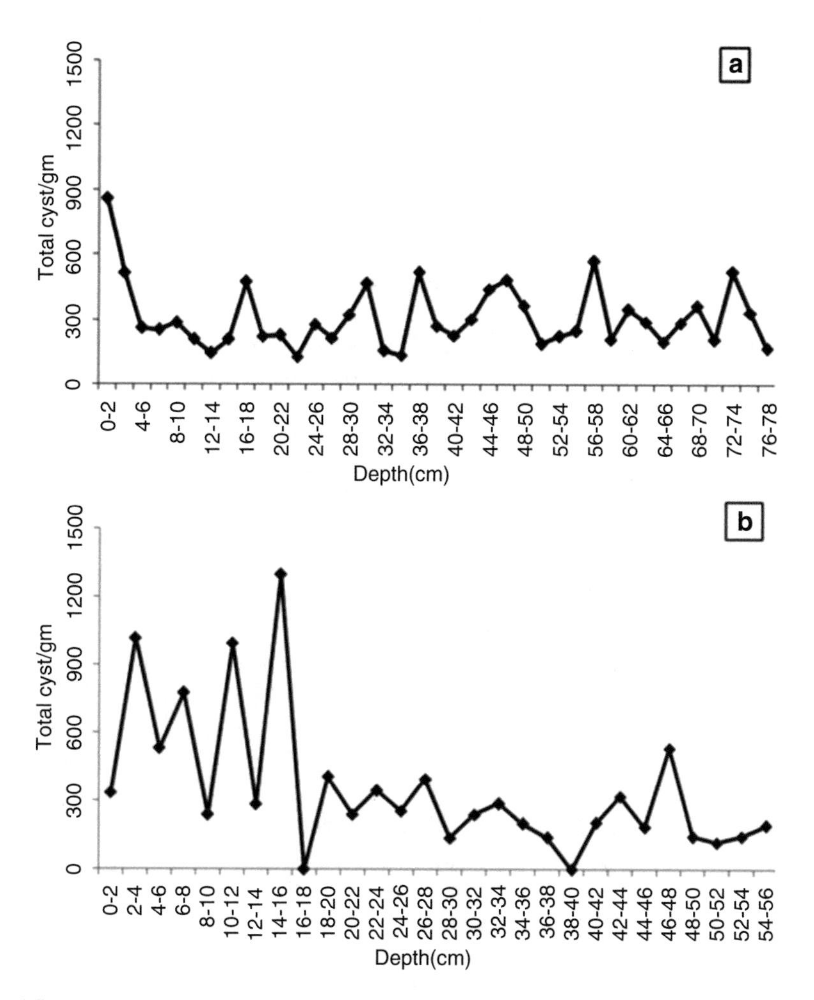

Fig. 1.5 Vertical distribution of total dinoflagellate cysts per gm sediment in cores **A** and B in the Eastern Harbour of Alexandria, Egypt

Table 1.2 Example of vegetative dinoflagellate species recorded in previous works and their occurrence in the core sediment as observed by Ismael et al. (2014)

Vegetative species	Bloom density range (celll^{-1})[30-31-32]	Cyst density (cyst gm^{-1})
Alexandrium minutum	1.7–26 × 10^6	13–267
Peridinium [S20] quinquecorne	4 × 10^3	< 58
Scrippsiella trochoidea	1.5–6 × 10^5	2–59

For interpretation of the distribution patterns of resting cysts, it is necessary to describe the sediment characteristics of the investigated cores. According to sediment analysis, the sediments vary from coarse silt to medium sand (4.8 Ø–1.7 Ø in core A and 4.07 Ø–1.13 Ø in core B). Organic content varied between 0.09% and 3.58% (0.09%–3.58% in core A and 0.2%–2.26% in core B).

1.3.1.4 Heterotrophic Cyst Assemblage and Composition

During the present study, six genera with thirteen heterotrophic species were identified from the two core sediments. Total heterotrophic cyst ranged from 12 to 820 cyst [gm^{-1}] [s8] wet wt representing 9% and 63%, respectively. Total heterotrophic species showed the same trend as the total number of cysts. In core the total number of heterotrophic species does not exceed 300 cyst [gm^{-1}] [s10] wet wt, except in layers 0–2, 16–18, 56–58 and 72–47 cm, where it reached to 478 cyst [gm^{-1}] wet wt (Fig. 1.6). Core B showed two layers: the lower layer from 18 cm to the bottom of the core, where the heterotrophic cysts do not exceed 209 cyst [gm^{-1}] [s11] wet, and the upper layer from 18 cm to the core surface, where the total heterotrophic cyst increased to 820 cyst gm^{-1} wet.

On the other hand, *Protoperidinium* spp., dominant among the heterotrophic species in the two sediment cores, ranged from 8 to 568 cyst [gm^{-1}] [s12] wet wt (Fig. 1.7). In core A, they represent 66.7%, while in core B, they represent 44.5% of total numbers of cysts. Sediment layers containing blooms of *Protoperidinium* spp. were characterized by nutrients rich in TP (1.6 × 103[S13]–7.4 × 103 [S14] ppm) and TN (2.1 × 103 [S15]–4.5 × 103 [S16] ppm) with an increase in diatom productivity (1.3 × 103 [S17]–2.7[S18] × 103 cell [g^{-1}] [S19]).

Three *Spiniferites* species were recorded from the core sediments in the E.H. during the present study: *Gonyaulax scrippsae*, *G. spinifera* and *G. verior*. These species were of higher abundance in core A at layers of increasing P concentration: 10–12 cm, 22–24 cm, 30–32 cm and 42–44 cm (Fig. 1.8).

1.3.1.5 Discussion

This study demonstrates trends of dinoflagellate cysts based on species composition, abundance and distribution along the core sediments in the Eastern Harbour of Alexandria. Our results document the composition of the assemblages that are dominated by *Brigantedinium* spp. (*Protoperidinium* spp.) and *Dubridiniu* spp. (*Diplopsalis* spp.). These taxa are commonly found in coastal and estuarine systems of temperate climate zone (Dale 1996). Distribution pattern of dinoflagellate cysts seems to be functions of sediment grain size and nutrient concentrations. A change in grain size is accompanied by change in the cyst abundance. Although Ismael and Khadr (2003) showed that there is no relationship between the sediment texture and the cyst distribution in the cores, the present study showed that cysts in core B tend to be more abundant in fine sand-size sediments with significant positive correlation

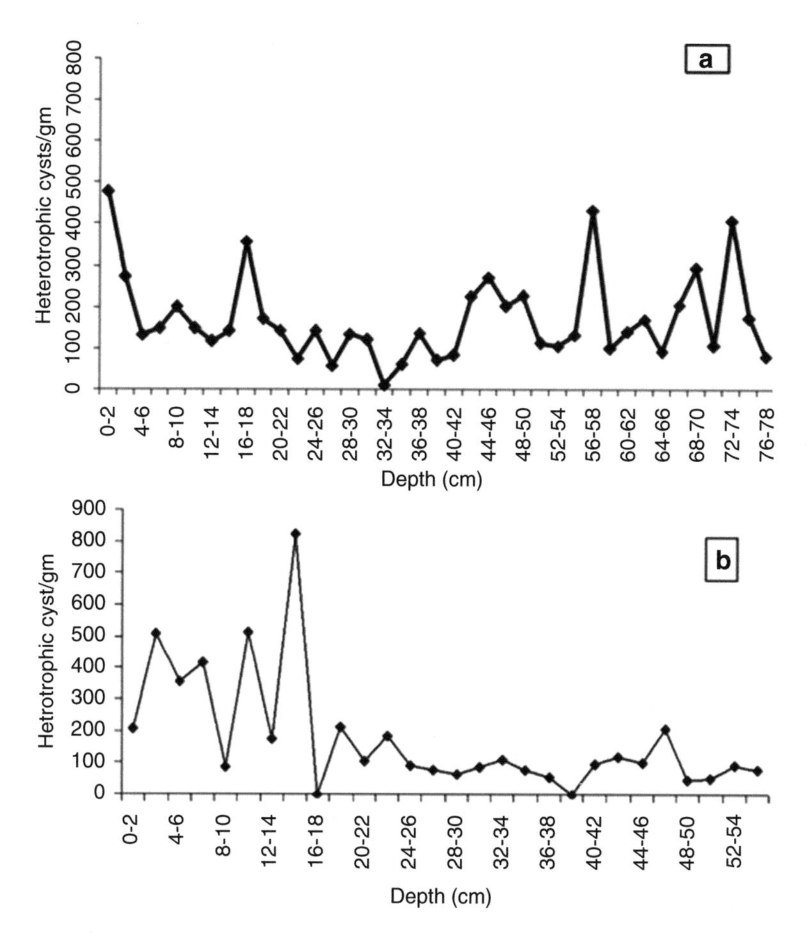

Fig. 1.6 Vertical distribution of heterotrophic dinoflagellate cysts gm^{-1} in cores **A** and **B** in the Eastern Harbour of Alexandria, Egypt

with Mz ($P \leq 0.05$, $r = 0.44$). Maximum concentration of dinocyst was recorded at very fine sand, while lower abundance was recorded in medium sand and coarse silt sediments. On the other hand, no significant correlation between total number of cysts and OM% was found along the two cores (Table 1.3).

Identification of past, present and future cultural eutrophication is one of the critical requirements for the safe and sustainable management of coastal waters. Understanding past occurrences helps to develop information that is crucial for the timely identification or prevention of present and future eutrophication. In past cases where eutrophication resulted in reduced water quality (e.g. lowering of dissolved oxygen levels), this offers possibilities for retrospectively estimating the critical threshold levels of nutrient loading for the ecosystems affected (Dale 2009).

The study of dinoflagellate cyst assemblage is useful to interpret the environmental conditions of the water column. An increase in the proportion of cysts of

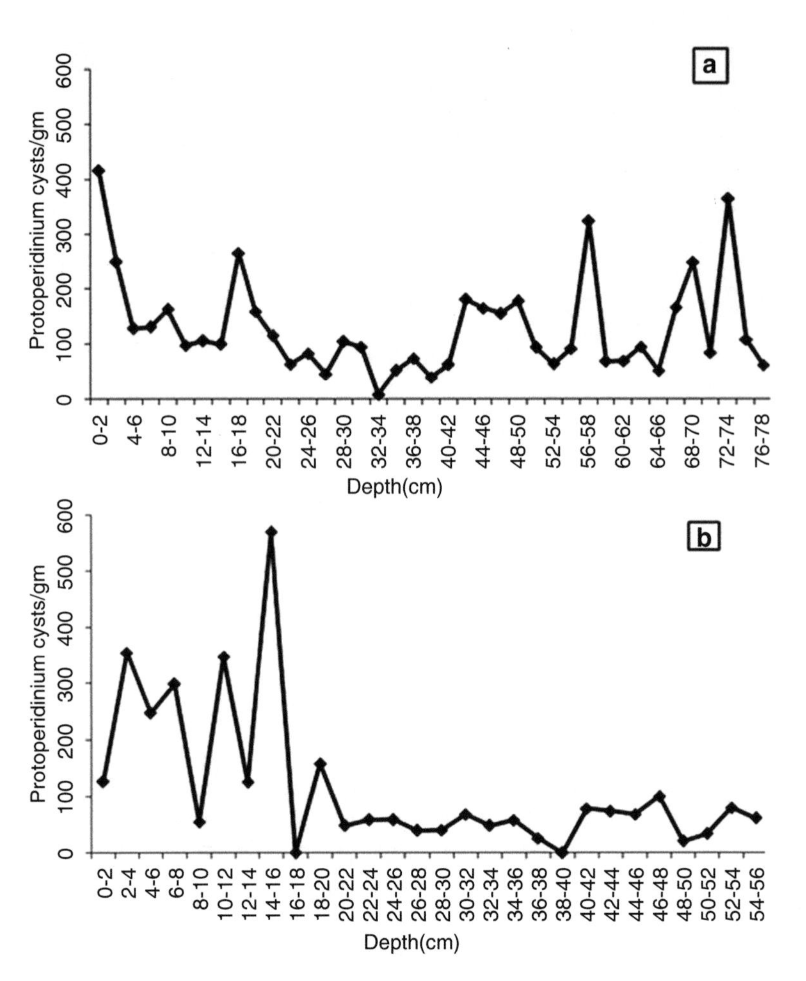

Fig. 1.7 Vertical distribution of *Protoperidinium* spp. gm^{-1} in cores **A** and B in the Eastern Harbour of Alexandria, Egypt

heterotrophic dinoflagellate (HTD) has been suggested as a signal of eutrophication and industrial pollution (Saetre et al. 1997; Matsuoka 1999; Matsuoka 2001; Pospelova and Head 2002). Matsuoka (1999) suggested that the increase in diatom production is the main cause of increasing heterotrophic dinoflagellate. The total heterotrophic cysts showed a significant positive correlation with diatoms, sedimentary TP and TN at $P \leq 0.05$, $r = 0.4$, 0.32 and 0.33, respectively. The results indicate that increasing in diatoms and nutrient concentrations are the main causative factors for increasing HTD in the Eastern Harbor.

In addition, the high percentages of *Protoperidinium* spp. are recorded in areas characterized by high nutrient inputs, low salinity in surface water and high primary productivity, including diatoms (De Vernal et al. 1998). Matsuoka (1999) and Matsuoka et al. (2003) pointed that the slightly higher concentration of heterotro-

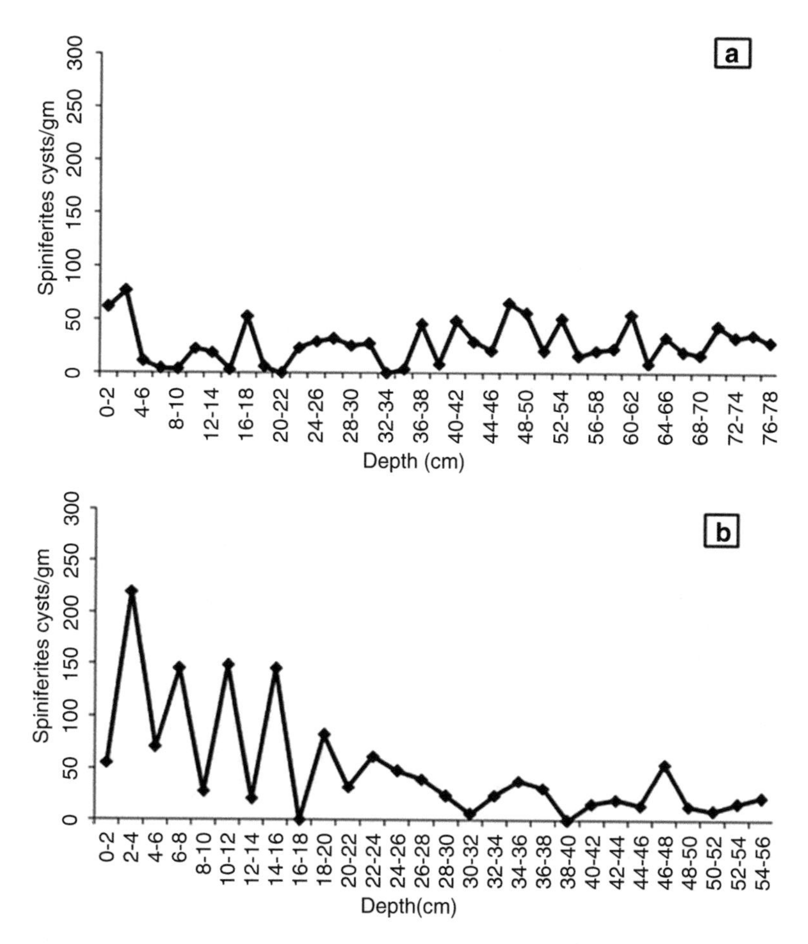

Fig. 1.8 Vertical distribution of *Spiniferites* spp. gm^{-1} in cores A and B in the Eastern Harbour of Alexandria, Egypt

phic *Protoperidinioids* suggests that the waters in the study area are eutrophic. This is the same signal during the case study as genus *Protoperidinium* in core B showed a significant positive correlation with diatoms, sedimentary TN and TP at $P \leq 0.05$, $r = 0.86$, 0.65 and 0.58, respectively. Their positive correlation with TN and TP supports the use of this group as an indicator of nutrient enrichment which is in agreement with the work of Pospelova and Kim (2010).

Although Norwegian studies have provided evidence that some cyst taxa may show a positive response to cultural eutrophication, in some fjords the number of *Lingulodinium machaerophorum* (*Lingodinium polyedra*) increased with the increasing nutrient enrichment. In other fjords an increase of the relative proportion of cysts of the heterotrophic species *Selenopemphix quanta* (*Protoperidinium conicum*) was also reported (Thorsen and Dale 1997). The cyst of *Lingodinium polyedra* in core A was recorded at layers of low TN and TP concentration with negative cor-

Table 1.3 Correlation matrix of total dinoflagellate cysts, *Protoperidinium* cysts, *Lingulodinium* cysts and *Spiniferites* cysts against MzØ, O.M. %, sedimentary TN, TP and diatom standing crop in cores A and B at 0.05 level

Parameters	Mz Ø	OM%	TN (ppm)	TP (ppm)	Diatoms (Cell.gm^{-1})
Core A					
Total cyst gm^{-1}	0.15	−0.09	0.01	−0.2	0.42
Protoperidinium gm^{-1}	−0.08	−0.25	−0.14	−0.08	0.13
Lingulodinium gm^{-1}	0.1	−0.03	−0.04	0.11	0.4
Spiniferites gm^{-1}	0.17	−0.19	−0.09	−0.27	0.2
Core B					
Total cyst gm^{-1}	0.44	0.3	0.62	0.53	0.87
Protoperidinium gm^{-1}	0.47	0.29	0.65	0.58	0.86
Lingulodinium gm^{-1}	0.29	0.5	0.25	−0.03	0.07
Spiniferites gm^{-1}	0.53	0.46	0.65	0.49	0.85

relation with sedimentary TP and TN concentrations. This can be compatible with the Tokyo Bay and Massachusetts Bay (Matsuoka 2001), where cysts of *Lingulodinium machaerophorum* and heterotrophic *Selenopemphix quanta* (*Protoperidinium conicum*) tend to have negative response to eutrophication and inorganic pollution. Cysts of the two species decreased over a period of increasing nutrient enrichment. Previous plankton records in the Eastern Harbour (Ismael 1993) showed the same trend where *L. polydra* was recorded in the water column at high salinity (38.7‰) and low nutrient concentrations (11.3 μg at N l^{-1} and 0.15 μg at P l^{-1}).

On the other hand, individual taxa respond to water quality changes in different ways (Pospelova and Chmura 1998). Cysts of *Spiniferites* spp. (*Gonyaulax* sp.) seem to be tolerant of extreme environmental conditions as they are the most abundant in cyst assemblages from low salinity and highly eutrophic conditions (Pospelova and Head 2002). Ismael et al. (2014) observed that there is significant positive correlation between the abundance of *Spiniferites* (*Gonyaulax*) assemblage and TN and TP in core sediment at $P \leq 0.05$ and $r = 0.27$. This is compatible with the earlier work (Ismael 1993), where *G. spinifera* appeared at low salinity (36.8‰) and high TN and TP concentrations (74.79 μg at Nl^{-1} and 8.01 μg at Pl^{-1}, respectively).

1.4 Hypoxic Condition

This is a typical stressed condition due to depletion of oxygen (low oxygen) encountered near inhabited coastal regions due to the release of excessive quantities of nutrients. The amount of oxygen in any water body varies naturally, over seasons and over time, due to a balance between oxygen input from the atmosphere and certain biological and chemical processes. Some of the processes produce oxygen, while others consume it. Water column stratification is the potential natural cause of hypoxia which occurs when less dense freshwater from an estuary mixes with heavier seawater. Limited vertical mixing between the water 'layers' restricts the supply of oxygen from surface waters to more saline bottom waters, leading to hypoxic conditions in bottom habitats.

Large areas of ocean eutrophication are associated with extensive hypoxic (*hypo = under, oxic = oxygen*) dead zones of oxygen-poor water that often occur near the mouths of major rivers after large spring runoff. When rivers deliver excess nutrients to the ocean, they feed profuse, widespread blooms of algae that later die and decompose, robbing the water of oxygen. Oxygen levels within these dead zones drop from above 5.0 parts per million (ppm) to below 2.0, which is lower than most marine animals can tolerate. Some of the more mobile marine organisms can free the area, but it suffocates and kills many bottom-dwelling organisms that can swim or can crawl to other areas. Low oxygen levels have also been shown to limit the growth and reproduction of marine life.

A dead zone is an area with frequent, persistent anoxic conditions; nevertheless they are very much alive with microorganisms. When algae die, decomposition consumes vast amounts of oxygen, creating an anoxic dead zone that affects a multitude of marine life. Global Environment Outlook Yearbook (*GEO Yearbook 2003*) reported 146 dead zones in the world's oceans where marine life could not be supported due to depleted oxygen levels. Some of these were as small as a square kilometre (0.4 mi^2), but the largest dead zone covered 70,000 square kilometres (27,000 mi^2). A 2008 study documented ~ 405 dead zones worldwide (Diaz and Rosenberg 2008); nevertheless the Gulf's dead zone is the largest in the Western Hemisphere. To control the spread of the Gulf's dead zone includes the following measures: (1) controlling nutrient runoff from agriculture, (2) preserving and utilizing wetland that filter runoff before it enters the Gulf, (3) planting buffer strips of trees and grasses between farm fields and streams, (4) altering the times when fertilizers are applied, (5) improving crop rotation and (6) enforcing existing clean water regulations. In this way, scientists, land planners and policymakers are working to institute an action plan that would shrink the yearly Gulf of Mexico dead zones.

One of the most prominent dead zones is the one that forms each summer near the mouth of the Mississippi River in the northern Gulf of Mexico of Louisiana. In 2002, in fact, it reached a record size of 22,000 square kilometres (8500 square miles). A report revealed in August 2017 that the US meat industry is responsible for the largest-ever dead zone in the Gulf of Mexico (Oliver 2017). Smaller dead zones have occurred each summer in the region for decades, but they dramatically increased in size after the record-breaking 1993 Midwest floods. Other notable dead zones include the surrounding of the outfall of the Mississippi River, the coastal regions of the Pacific Northwest and the Elizabeth River in Virginia Beach, all of which have been shown to be recurring events over the last several years.

Dead zones may first manifest in coastal ecosystems as a deposition of organic material which promotes microbial growth and respiration resulting a greater biochemical oxygen demand (BOD) (Diaz and Rosenberg 2008). This can lead to oxygen depletion if accompanied by stratification of the water column. Eventually extreme hypoxia progresses resulting mass mortality of marine biota if the process of eutrophication continues (Diaz and Rosenberg 2008). Extreme hypoxia is generally regarded as when the oxygen concentration reaches <2 ml l^{-1} at which point benthic infauna exhibits stress-related behaviour, such as abandonment of burrows with significant mortality occurring at levels of <0.5 ml l^{-1} (Diaz and Rosenberg 2008). However, studies of oxygen thresholds indicate that for sensitive taxa, such as fish or crustaceans, lack of oxygen may be lethal at concentrations well above 2 ml^{-1} and there is considerable variability among taxa in sensitivity to hypoxia (Vaquer-Sunyer and Duarte 2008). Progressive accumulation of organic material and nutrients can result in seasonal hypoxia with regular mortality events of marine organisms. In the worst cases, where nutrients continue to accumulate in the system over several years, the hypoxic zone will expand, and anoxia may be established accompanied by release of H_2S by microbial communities (Diaz and Rosenberg 2008). Ecological thresholds of dissolved oxygen (DO) and sedimentary hydrogen sulphide (H_2S) for benthic organisms were investigated by Kanaya et al. (2018) in a highly eutrophic canal in Tokyo Bay, Japan. They had observed that dominant opportunistic macrozoobenthos, e.g. polychaetes and amphipods, were eliminated under low DO and high H_2S conditions.

This chronic environmental problem mainly evolved as a result of multiple human- and nature-induced stresses. The former is concerned with significant enrichment of inorganic nutrients (recognized as eutrophication), especially of nitrogen and phosphorous. These are originated from multiple sources, such as urban, agricultural and sewage effluents, fossil-fuel burning, urban land use, coastal development and waste water treatment effluents. Natural causes are related to coastal upwelling and changes in wind as well as water circulation patterns.

Additionally, natural oceanographic phenomena can cause deoxygenation of parts of the water column. For example, enclosed bodies of water, such as fjords or the Black Sea, have shallow sills at their entrances, causing water to be stagnant there for a long time. The eastern tropical Pacific Ocean and northern Indian Ocean have lowered oxygen concentrations which are thought to be in regions where there is minimal circulation to replace the oxygen that is consumed (Pickard and Emery 1982). These areas are also known as oxygen minimum zones (OMZ). In many cases, OMZs are permanent or semipermanent areas.

The effects of hypoxia on the microbial food web are not quite distinct. Heterotrophic bacteria and protists are responsible for reducing oxygen levels in these waters, and they are generally abundant and grow well in hypoxic/anoxic waters (Fenchel et al. 1990; Bastviken and Tranvik 2001). Hence, rather than being 'dead zone', hypoxic/anoxic waters in coastal regions and estuaries are quiet alive, and microbial food webs completely dominate biological activity in this adverse environment.

1.4.1 Hypoxia and Climate Change

Coastal hypoxia (depletion of oxygen in bottom waters) has become an emerging global problem causing severe ecological hazards, such as loss of benthic habitat and faunal diversity. This should be recognized as a potential threat to marine biodiversity. This phenomenon evolves when the oxygen supply is not sufficient enough to comply with the oxygen consumption for a long period of time. Hypoxia is recognized as a natural phenomenon in fjords or basins with limited water circulation (e.g. Black Sea), or in shelf regions subject to the upwelling (i.e. upward movement of deeper waters) of oxygen-poor and nutrient-rich subsurface water (e.g. Arabian Sea, Northeast Pacific). Besides natural hypoxia, a global increase in the intensity, frequency and spatial extent of coastal hypoxia is also evident related to intensive anthropogenic activities. Prevalence of coastal hypoxia phenomenon is steadily increasing with a 5.5% per year over the past three decades in the number of coastal sites with such specific stresses. Changes in ocean circulation triggered by ongoing climate change could also add or magnify other causes of oxygen reductions in the ocean (Mora et al. 2013).

Climate change and enrichment of nutrient loadings are the two important potential drivers of global change which are related to the increase in frequency of coastal hypoxia. Model simulations were performed by Meire et al. (2013) for investigating the relative influence of climate change and nutrient run off on oxygen dynamics in bottom waters in the Oyster Grounds, a potentially hypoxic area in the Central North Sea, which is characterized by summer stratification. It is a large, roughly circular, depression in the central North Sea with its centre at 54° 30′ N and 4° 30′ E. The low oxygen level is developed in the area due to the existing ideal conditions such as (1) shallow topography and stratified nature, (2) potentially high nutrient concentrations and (3) suitable sedimentary structure. Through one-dimensional ecosystem model, they had pointed out that risk of hypoxia will be greater due to climatologically changing condition.

The oxygen dynamics is especially sensitive to increasing temperature as the oxygen solubility decreases in warmer water, while oxygen consumption rates increase due to respiration. Increase in sea surface temperature (SST) may trigger the negative consequences of eutrophication, especially in stratified estuaries. Intense stratification can be considered as the potential agent linked for the decrease in bottom water oxygen concentration. In addition to climate change effects, change in nutrient runoff loadings also acts as the factor on the bottom water oxygenation. The frequency of hypoxic events will be much increased due to greater influx of inorganic nutrients. Hence eutrophication management is urgently required to counteract the effect of rising temperature.

The consequences of global warming as hypoxia in the North Sea-Baltic Sea were investigated by Hansen and Bendtsen (2009) for 3 consecutive years (2001–2003), by reproducing observation of spatiotemporal distribution of oxygen in this transition zone. It was revealed that the area was affected by severe hypoxia, as defined by oxygen concentration less than 60 μM O_2.

The temperature effect may to some oxygen be compensated by reducing the nutrient concentration in the surface layers. In order to keep oxygen concentration at the present-day level, reductions in the nutrient concentration of up to 30% are required. Similar situations are expected in several stratified estuaries, and hypoxia may cause the most significant adverse effect for global warming on coastal biodiversity.

1.5 Principal Groups of Algal Bloom Causative Organisms

1.5.1 Diatoms (Phylum, Chrysophyta; Class, Bacillariophyceae)

- Unicellular algae, characterized with cells contained in the shell or frustules and composed of amorphous silica (Sio_2. nH_2O): silica with skeleton forms up to 4–50% of the dry weight, thus required by diatoms in the relatively large amount.
- Frustule is similar in structure to a microscopic pill box, composed of heterogeneous values epitheca (the larger valve) and hypotheca (the smaller one).
- A key component contributing substantial part of the phytoplankton community in waters as they are cosmopolitan in distribution from the temperate to the polar zone (Arctic and Antarctic) and have dominant role in global biogeochemical cycle. In addition, some diatoms (e.g. *Pseudo-nitzschia*) can synthesize the potential neurotoxins responsible for amnesic shellfish poisoning, discussed in Chap. 3.
- Divided into mainly pinnate diatoms and characterized by the following features: commonly bilaterally symmetrical, elliptical and benthic and planktonic centric diatoms provided with radial symmetrical frustules.
- Cell size range ~2 μm to >100 μm and may occur as single larger chains or other forms of aggregates in which individual cells are held together by mucilaginous threads or spines.
- Examples: *Coscinodiscus* sp., *Chaetoceros* sp. and *Skeletonema* sp.

1.5.2 Dinoflagellates (Phylum, Pyrrophyta; Class, Dinophyceae)

- Biflagellate, unicellular and highly evolved microalgae with a size of 5 μm–2 mm, often dominate the tropical and subtropical phytoplankton community and also may dominate the summer and autumn phytoplankton of the temperate and boreal zone.
- Motile cells provided with two unequal flagella: one flagellum is located in the groove that divided the cell into two subequal parts. The other flagellum is oriented perpendicularly to the transverse flagellum and extends posteriorly.

- External walls do not contain silica but may be covered with a series of continuous cellulose, giving them the appearance of spinning armed helmets. The pattern of theca is usually the most important diagnostic feature of the species.
- Morphologically and functionally diverse and constitute the key component of marine phytoplankton and play significant role in food webs, displaying a wide variety of trophic strategies such as autotrophy, heterotrophy and mixotrophy.
- Dinoflagellates are characterized by large genome size ranging from 1.5 to 185 Gbp, based on next-generation sequencing (NGS) which is used for assessing phytoplankton diversity (Casabianca et al. 2017).
- In theca-bearing species, cell wall is divided into a number of separate cellulose plates that are equipped with pores and/or small spines, e.g. *Ceratium*, *Gonyaulax*, *Dinophysis* sp.

1.5.3 Blue-Green Cyanobacteria (CB) (Members of the Group Cyanophyceae)

- Widespread prokaryotic organisms occur both as free-living forms and symbiotically associated with marine benthic plants and animals.
- Becoming more numerous, widespread and persistent in nutrient-enriched estuarine and coastal waters worldwide.
- Photosynthetic but many also are capable of nitrogen fixation, the conversion of gaseous nitrogen into ammonium ion and synthesis of amino acids.
- Tend to favour environments where there is local absence of oxygen because nitrogen fixation is favoured under this stressed condition.
- Gulf of Finland is known for its heavy late summer cyanobacteria blooms, mainly caused by *Nodularia spumigena* and *Dolichospermum* spp.
- The interaction between cyanobacteria and zooplankton is quite complex.
- Engström-Öst et al. (2015) demonstrated that *Nodularia* causing cyanobacteria blooms had negative impact on the calanoid copepod *Acartia* (the dominant genus in global coastal waters) reproductive output in the Baltic Sea, western Gulf of Finland. The copepod population dynamics, too, was negatively affected by such blooms.
- The bloom of CB is increasing both locally (Suikkanen et al. 2013) and globally (Pearl and Otten 2013), and their occurrences seem to continue an increasing trend in marine and estuarine environments in the future (reviewed by O'Neil 2012). They are favoured from climate change in diverse means (Pearl and Huisman 2009).

List of algal bloom causative organisms recorded from different coastal waters of India has been depicted in Table 1.4. The scanning electron microscope (SEM) and microphotographs of some phytoplankton have been depicted in Figs. 1.9 and 1.10, respectively.

Table 1.4 List of algal bloom causative organisms recorded from different coastal waters of India

Sr. no.	Causative organism	Place of occurrence	Year	Season	Effect	Cell count (CC), Chl *a*	References
A. Bacillariophyceae							
1	*Ditylum* sp., *Thalassiosira* sp.	Malabar coast	1922	PrM	–	–	Hornell and Nayudu (1923)
2	*Fragilaria oceanica*	Off Kaikani, Mangalore	1972	SWM	–	CC: 36.8 × 10^6 cells L^{-1}, Chl a: 123.5 lg L^{-1}	Devassy (1974)
3	*Nitzschia sigma*, *Skeletonema costatum*	Cochin backwaters, Kerala	1970	PrM	–	CC: 1.4 × 106 cells L^{-1}	Devassy and Bhattathiri (1974)
4	*Skeletonema costatum*	Dharamtar creek, Mumbai	1985	PoM	–	CC: 9 × 105 cells L^{-1} Chl *a*: 6.4 mg m^{-3}	Tiwari and Nair (1998)
5	*Coscinodiscus asteromphalus* var. *centralis*	Off Kodikkal-Calicut, Kerala coast	1905	SWM	Brownish-red discolouration of water	CC: 7 × 106 cells L^{-1} Chl *a*: 206.5 mg m^{-3}	Padmakumar et al. (2007)
6	*Gonyaulax polygramma*	Off Cochin, Kerala coast	1963	PoM	Non-toxic but virtual exclusion of zooplankton	CC: 10.8 × 106 cells L^{-1}	Prakash and Sarma (1964)
7	*Nitzschia miliaris*	Off Quilon, Kerala	1976	SWM	Red colouration of water	CC: 2.4–4.1 × 105 m^{-3}	Venugopal et al. (1979)
8	*Alexandrium tamiyavanichi*	Kumble estuary, Mangalore coast	1983	PrM	Reports of PSP, one death reported and several hospitalized after consumption of clams *Meretrix casta*	–	Karunasagar et al. (1984)
9	*Nitzschia miliaris*	Mandovi and Zuari estuaries; coastal waters of Goa	1987	PrM	Green colouration of water, fish catch decreased	CC: 0.2–5.1 × 104 cells L^{-1}, Chl a: 16.7 mg m^{-3}	Devassy and Nair (1987)
10	*Nitzschia miliaris*	Mangalore	1987	PoM	Intense green colouration of water	CC: 1.6 × 104–7.6 9109 m^{-3}	Katti et al. (1988)

11	*Gymnodiniumnagasakiense*	Brackish water fish farm at Kodi, Karnataka	1989	PoM	Red colouration of water and fish mortality	CC: 4 × 108 cells L^{-1}	Karunasagar (1993)
12	*Nitzschia miliaris*	Off Mangalore	1993	PrM	Increased proportion of *Moraxella*-like bacteria with bloom	CC: 1.6 × 106 cells m^{-3}	Nayak et al. (2000)
13	*Nitzschia miliaris*	Off south Thiruvananthapuram, Kerala coast	2004	SWM	Red discolouration of water	CC: 9 × 105 cells L^{-1}	Sahayak et al. (2005)
14	*Nitzschia miliaris*	Off Goa	2008	PoM	No fish kills	CC: 2 × 104 cells L^{-1}	Sanilkumar et al. (2009)
15	*Coscinodiscus marina*	Off Kochi, Kerala	2009	SWM	Rusty brownish-red discolouration of water	CC: 1.59 × 107 cells L^{-1}, Chl a: 8.3 lg L^{-1}	Padmakumar et al. (2011)
16	*Asterionellopsis glacialis*	Kalpakkam, Tamil Nadu	2015	NEM	Dense and created brownish-coloured patches near the surf zone	CC: 5.63 × 107 cells l^{-1}, Chl a: 15.99 mg m^{-3}	Sahu and Mukhopadhyay (2015)
B. Dinophyceae							
17	*Noctiluca* sp.	Cochin-Calicut, off Kerala coast	1998	SWM	Oxygen depletion resulted in severe mortality of fish	–	Nagvi et al. (1998)
18	*Cochlodinium polykrikoides*	Off Goa	2001	PoM	Fish mortality coincided with bloom occurrence	–	O'Herald (2001)
19	*Noctiluca scintillans*	Off Goa to Porbandar, (Gujarat) coast	2003	PrM	Green colouration of water	CC: 0.6–25.42 × 102 cells L^{-1}	PrabhuMatondkar et al. (2004)
20	*Karenia mikimotoi*	Kerala coast	2004	SWM	Mass mortality of fish	CC: 9.0 × 104 cells L^{-1}, Chl a: 8.8–721 mg m^{-3}	Iyer et al. (2008)

(continued)

Table 1.4 (continued)

Sr. no.	Causative organism	Place of occurrence	Year	Season	Effect	Cell count (CC), Chl *a*	References
21	*Noctiluca miliaris*	Off Gujarat	2007	PrM	Deep-green colouration of water	CC: 4 × 106 cells m^{-3} Chl *a*: 21.9 mg m^{-3}	Padmakumar et al. (2008b)
22	*Protoperidinium* sp.	Mangalore coast	2008	PoM	No fish kills	CC: 8.1 × 108 cells L^{-1}, Chl a: 12.3 lg L^{-1}	Sanilkumar et al. (2009)
23	*Noctiluca scintillans*	Off Kochi, Kerala	2008	SWM	Brick red discolouration of water, no fish mortality	CC: 5 × 108 cells L^{-1}	Padmakumar et al. (2008b)
24	*Karenia mikimotoi*	Cochin barmouth, Kerala	2009	PoM	Intense brownish colouration to water	CC: 7.0–15.5 × 106 cells L^{-1}, Chl a: 24–85.8 mg m^{-3}	Madhu et al. (2011)
25	*Noctiluca scintillans*	Rushikulya river, south Orissa coast	2005	PrM	Red discolouration of water, oxygen depletion	CC: 2.83 × 105 cells L^{-1}	Mohanty et al. (2007)
26	*Noctiluca scintillans*	Gulf of Mannar, Orissa	2008	PoM	Deep-green colouration of water. Corals got bleached, due to lack of oxygen. Many fishes and sea animals died	CC: 13.5 × 105 cells L^{-1}, Chl a: 116 mg m^{-3}	Gopakumar et al. (2009)
C. Cyanophyceae							
27	*Trichodesmium erythraeum*	Mangalore-Quilon	2005	PrM	Discolouration of water, no mortality reported	CC: 0.5 × 105 filaments ml^{-1}	Anoop et al. (2007)
28	*Microcystis aeruginosa*	Chalakudy River in Central Kerala	2008	PrM	Discolouration of water. Itching, irritation of the skin among local people	CC: 4 × 108 cells m^{-3} Chl *a*: 23.4 mg m^{-1}	Padmakumar et al. (2008a)
29	*Trichodesmium erythraeum*	Off Kollam, Kochi and Kannur, Kerala coast	2009	Onset of SWM	Brown discolouration of water	CC: 1.14 × 106, 1.97 × 106, 1.51 × 106 filaments L^{-1}	Padmakumar et al. (2010)

30	*Trichodesmium erythraeum*	Tamil Nadu coast	2007	Winter monsoon	Discolouration of surface water	CC: 13275 cells L^{-1}, Chl a: 2.0–7.5 μgL^{-1}	Rajasekar et al. (2010)
31	*Trichodesmium erythraeum*	Kalpakkam, Tamil Nadu	2007	PrM	Yellowish-green colouration of water. No fish mortality reported	CC: 4.1 × 106 cells L^{-1}	Satpathy et al. (2007)
32	*Trichodesmium erythraeum*	Arabian Sea, west coast	2009	Spring	Dark brown colouration of water	CC: 25000 cells L^{-1}, Chl a: 67 mg m^{-3}	Roy et al. (2011)
33	*Trichodesmium erythraeum*	Mandapam and Keelakarai, Tamil Nadu	2008	PoM	Mortality of several fishes and shellfishes	–	Anantharaman et al. (2010)
34	*Microcystis aeruginosa*	Vellar estuary, Tamil Nadu	2009	NEM	Greenish-brown patch observed	CC: 37.6 × 103 colony L^{-1}, Chl a: 18.61 lg L^{-1}	Santhosh Kumar et al. (2010)

PrM pre-monsoon, *SWM* southwest monsoon, *PoM* post-monsoon, *NEM* northeast monsoon

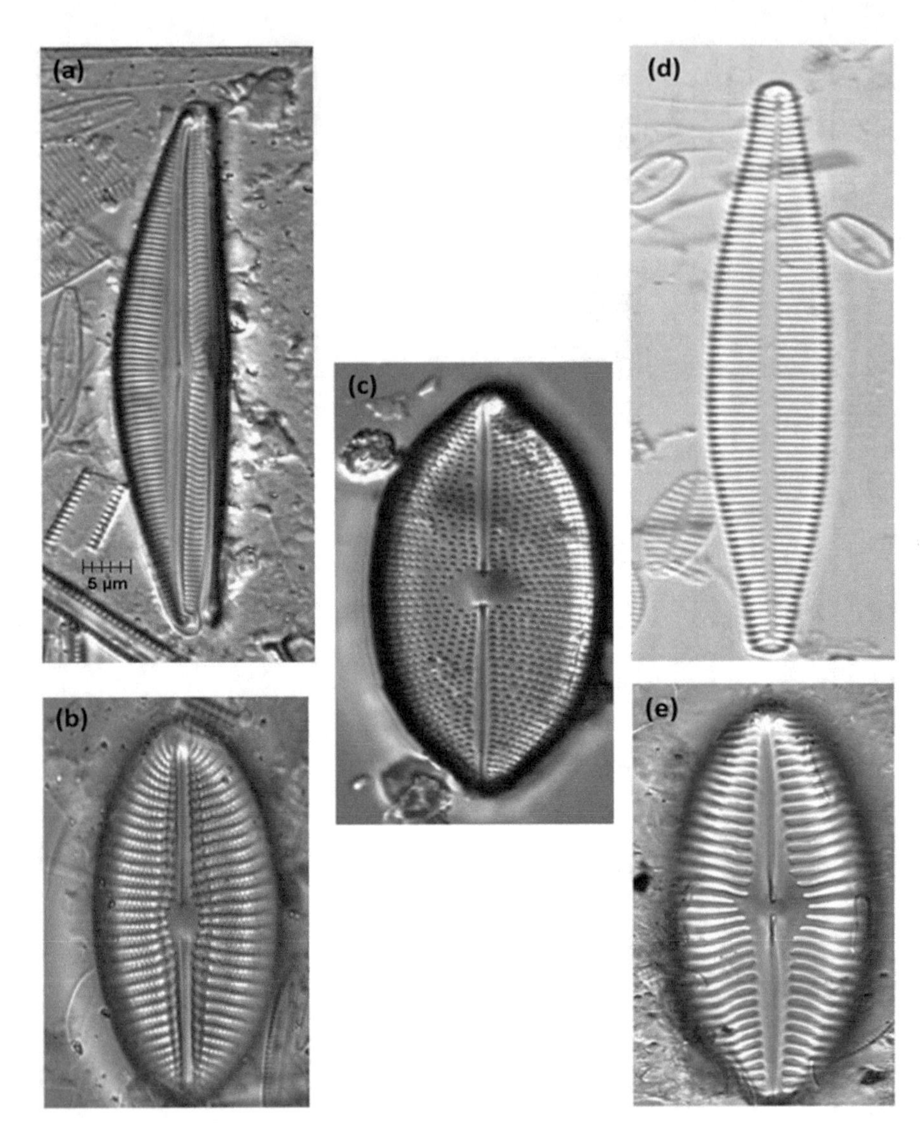

Fig. 1.9 SEM images of (**a**) *Amphora ventricosa,* (**b**) *Diploneis smithii*, (**c**) *Navicula marina*, (**d**) *Fragilaria* sp., (**e**) *Navicula yarrensis*. (*Courtesy*: Dr. Kathryn Taffs, Southern Cross University, Australia)

1.6 Remote Sensors in Algal Bloom Monitoring

Monitoring and assessing the quality of surface waters are critical for managing and improving its quality. Thus, in situ measurement and collection of water samples and their subsequent analyses in the laboratory are not sufficient to give a huge spatiotemporal view needed for water resource management. Hence, assessment of

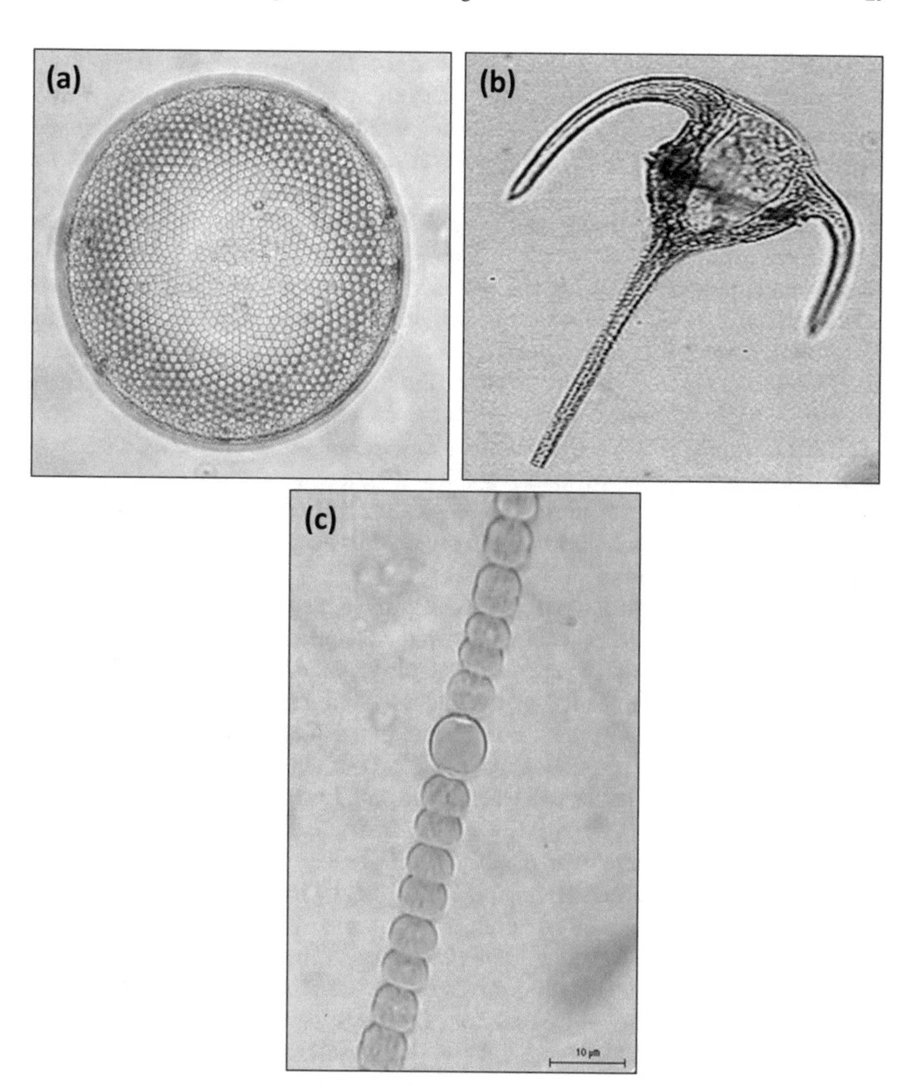

Fig. 1.10 Microphotographs of (**a**) *Coscinodiscus radiatus* (Bacillariophyceae), (**b**) *Ceratium tripos* (Dinoflagellate), (**c**) *Anabaena* sp. (Cyanophyceae) recorded from Sundarban mangrove wetland, West Bengal, India

water quality parameters in water bodies is one of the most scientifically relevant and commonly used applications of remote sensing. Remotely sensed data with high spatial resolution and frequent acquisition frequency offer solution to monitor variability of water quality parameters up to several times per year. Application of remotely sensed data allows to discriminate between water quality parameters and to develop a better understanding of light, water and substance interactions.

Remote sensing technologies provide a potential tool for frequent, synoptic water quality characteristics covering large areas (Lee et al. 2005). Satellites detect changes in the way the sea surface reflects light. These changes can be linked to concentrations of chlorophyll, showing where algae and other ocean plants are concentrated in the ocean. Chlorophyll *a* is the predominant type found in algae and cyanobacteria (blue-green algae), and its abundance is a good indicator of the amount of algae present in the waters.

Therefore, satellite imagery can be used to complement traditional monitoring techniques (Domingues et al. 2008; Gohin et al. 2008), offering a substantial improvement in terms of spatial and temporal coverage for the upper layers of the water column, especially when a densification of sampling efforts is necessary for monitoring purposes. In optically complex waters, such as coastal waters affected by river outflows (Morel and Prieur 1977), the standard and global algorithms used in clear or 'oceanic' waters present inaccuracies (e.g. Carder et al. 1999; Novoa et al. 2011). Inputs of suspended and dissolved substances through river discharges alter the optical properties of water, significantly affecting accurate chl a estimation (IOCCG 2000). As these substances are region and season specific, regionally parameterized empirical algorithms are required to accurately estimate chl a at a local scale (IOCCG 2000).

As previously discussed, input of nutrients from agricultural and urban runoff, or those produced by coastal upwelling, is the key factor for causing algal blooms in many estuaries and coastal waters. Algal blooms induce eutrophic conditions, depleting oxygen levels needed by organic life, limiting aquatic plant growth by reducing water transparency and producing toxins that can harm fish, benthic animals and humans. The magnitude and frequency of algal blooms have increased globally in recent decades, as evidenced from data from ocean colour sensors onboard satellites.

Satellite and airborne measurements of spectral reflectance (ocean colour) represent an effective way for monitoring phytoplankton by its proxy, chlorophyll a, the green pigment that is present in all algae. This article reviews the use of remote sensing techniques for detecting phytoplankton and mapping algal blooms. The ocean colour study has been initiated using the Landsat satellite data in the 1970s. Then exclusively the Nimbus Coastal Zone Colour Scanner (CZCS) mission started in 1978 and continued successfully till 1986 with exclusive data collection and utilization programmes globally. Later on improvement in the 1990s, ocean colour remote sensing started in terms of precision-based improved sensor-based satellite missions like Indo-German Collaboration IRS-P3 MOS, NASA's Sea-viewing Wide Field-of-View Sensor (SeaWiFS), Japan's OCTS and IRS-P4 Ocean Colour Monitor-OCM (Oceansat-1) – the first exclusive ocean colour satellite of India along with microwave radiometer MSMR. Now the major ongoing ocean colour missions are the moderate-resolution imaging spectroradiometer (MODIS) since 2002 and from India the Oceansat-2 comprising the OCM-2 and the scatterometer on board, launched into orbit on September 23, 2009. So, there has been completion of about 35 years of having ocean colour data sets. There are multiple potential applications of ocean colour datasets, such as chlorophyll concentration mapping, potential fishery zone (PFZ) mapping, suspended sediments, diffuse attenuation coefficient (Kd), ocean primary productivity, new primary production and colour dissolved organic matter (CDOM)/Yellow substances, etc. measurement.

The NASA coastal zone colour scanner (CZCS), operated from the satellite Nimbus 7 and launched in 1978, provided new insights into the complicated structure and dynamics of ocean water masses and their biological and chemical properties in time and space. Satellite imagery from the CZCS has revolutionized the view of primary productivity patterns in the ocean. Afterwards other ocean colour missions formulated, including the ADEOS/ocean colour temperature sensor (OCTS) in 1997, IRS-P3 MOS in 1996, NASA SeaWiFS mission in 1997, Indian missions IRS-P4 OCM/Oceansat-1 in 1999, Oceansat-2/OCM in 2009, NASA MODIS Aqua and Terra in 2002, ESA's MERIS in 2006, GOCI (geostationary platform-based mission) in 2010, NPP-VIIRS, HICO mission from ISS in 2011, recently mission Sentinel-3 in 2015, etc., are several sensors that produced very good-quality data sets for the ocean colour and many ocean applications. The Indian satellite Oceansat-3 is also planned to be launched during 2018. These sensors have used to monitor chlorophyll concentrations in surface waters, suspended sediment dynamics, primary productivity, fishery information and algal blooms and to produce images of coastal ecosystem dynamics, nutrient dynamics, nitrate modelling, etc.

1.6.1 Specifications of Some Remote Sensors

As the marine biological processes are closely knit with atmospheric and hydrospheric factors of physical and physico-chemical nature, almost all the sensors on board remote sensing platforms are to be considered as of marginal applicability. However, the sensors devised purely for marine biological applications are few and are those intended to detect water colour. The following account attempts to enumerate these sensing systems and state the relevant specifications of each.

1.6.1.1 Nimbus 7/CZCS

Nimbus 7 launched in 1978 by NASA carried an advanced payload of eight different meteorological, oceanographic and earth surface sensing instruments, and CZCS was among them. The orbit specifications of the satellite and features of CZCS are given in Table 1.5.

CZCS is a six-channel image scanner of high sensitivity, operating in visible, near-infrared and thermal-infrared regions of EMR. Its objective is to measure pigment concentration, sediment load and surface temperature in the sea. The salient features of CZCS are that the spectral bands at 443 and 670 nm are centred on the most intense absorption bands of chlorophyll, while a band at 550 nm is centred on the 'hinge point', the wavelength of minimum absorption. In spite of the termination of thermal channel in 1981, innumerable scenes of the oceans have been obtained and are being analysed, the world over.

Table 1.5 The orbit specifications of the satellite and features of CZCS: Nimbus 7

Orbit specifications		
Altitude	955 km	
Node	Ascending	
Revisit	6 days	
Orbit	Sun-synchronous	
Separation	26°	
Inclination	99.2°	
Equator crossing	Noon	
Orbit time	104 min	
Orbits/day	13.8	
CZCS spectral bands		
Spectral bands	Centre	Half-widths
Visible/near IR	443 nm	20 nm
	520 nm	20 nm
	550 nm	20 nm
	670 nm	20 nm
	750 nm	100 nm
Thermal IR	11.5 μm	2 μm
Pixel size	825 m	

Table 1.6 Basic data characteristics of CZCS II: (NOSS/CZCS) specifications

IFOV	1.13 mrad	
Pixel size	791 m	
Radiometric resolution	10 bit	
Spectral bands	**Centre**	**Half-widths**
Visible/near IR	440 nm	20 nm
	490 nm	20 nm
	560 nm	20 nm
	590 nm	20 nm
	650 nm	20 nm
	685 nm	20 nm
	765 nm	40 nm
	867 nm	45 nm
Thermal IR	10.3–11.3 μm	
	11.5–12.5 μm	

1.6.1.2 NOSS/CZCS

The National Oceanic Satellite System (NOSS) was intended as a 5-year operational demonstration of producing ocean data for use in near real time, based upon remote sensing from space-borne instruments. Among four instruments proposed for the system was CZCS, called CZCS II whose specifications are summarized in Table 1.6. One of the goals of NOSS mission was near instantaneous return and

Table 1.7 The orbital specifications of proposed NOAA satellite and the features of OCI

NOAA-J orbital parameters		
Altitude	870 km	
Orbital time	102.3 min	
Separation	25.587°	
Node	Ascending	
Inclination	98°	
Orbits/day	14	
Equator crossing	1.30 PM	
Revisit	8 days	
OCI specifications		
Spectral bands	**Centre**	**Half-widths**
Nominal	440 nm	20 nm
	520 nm	20 nm
	550 nm	
	670 nm (660)	
	765 nm	50 nm
	865 nm	50 nm
Experimental	490 nm	20 nm
	685 nm	20 nm
Tilt		0–20°

processing of global data on wind. Sea state and sea surface temperature with a nominal delay of about 3 h are required between observation and transfer of data. However, the programme was unfortunately deleted in 1981.

1.6.1.3 ERS-1/CZCS

In the wake of the deletion of NOSS and after removing another sensor called Ocean Colour Monitor (OCM) from the payload of ensuing satellite ERS-1 of ESA, it was considered to fly a CZCS. This CZCS was supposed to be an improved version of its counterpart on-board Nimbus, but, despite the general interest of biologists, the matter is no more pursued.

1.6.1.4 NOAA/CZCS

An advanced CZCS called ocean colour imager (OCI) as part of the operational satellite programme is being studied by NASA and NOAA. The orbital specifications of proposed NOAA satellite and the features of OCI are summarized in Table 1.7. In all, there is an addition of spectral channels in comparison to Nimbus CZCS, while in anticipation of advanced very high-resolution radiometer (AVHRR) data, the thermal channel is deleted. The OCI on-board NOAA satellite is expected to be in orbit sometime in 1986 or later.

1.6.1.5 Geostationary Ocean Colour Imager (GOCI)

GOCI, the world's first orbit satellite image sensor, was launched on June 27, 2010. It is one of the three payloads on the Communication, Ocean and Meteorological Satellite (COMS) provided with six visible bands (412, 443, 490, 555, 660 and 680 nm) and two near-infrared bands (745 and 865 nm). The observational coverage is 2500 km × 2500 km, centred on 36°N, 130°E. GOCI was developed to capture ocean colour data eight times each day; thus it can detect subtle changes in ocean environments around the Korean Peninsula (Ryu et al. 2012). The primary advantage of GOCI over other ocean colour satellite imagery is that it can obtain data every hour during the daytime, allowing ocean monitoring in near real time. GOCI can be effectively applied to monitor the temporal dynamics of the turbidity of coastal waters, i.e. sediment movements driven by the tidal cycle and short-term and long-term monitoring red tides. GDPS is specialized data processing software for GOCI data processing system, which was developed for real-time generation of various products.

The ocean data products that can be derived from the measurements are mainly the chlorophyll concentration, the optical diffuse attenuation coefficients, the concentration of dissolved organic material or yellow substance and the concentration of suspended particles in the photic zone of the sea. In operational oceanography, satellite-derived data products are used in conjunction with numerical models and in situ measurements to provide forecasting and now casting of the ocean state. Such information is of genuine interest for many categories of users.

Nutrient enrichment from agricultural and urban runoff, or those produced by coastal upwelling, is causing algal blooms in many estuaries and coastal waters. Algal blooms induce eutrophic conditions, depleting oxygen levels required for sustenance of life, limiting aquatic plant growth by reducing water transparency and producing toxins that can harm fish, benthic organisms and humans. The magnitude and frequency of phytoplankton blooms have increased globally in recent decades, as shown in data from ocean colour sensors on-board satellites. Satellite and airborne measurements of spectral reflectance (ocean colour) represent an effective way for monitoring phytoplankton by its proxy, i.e. chlorophyll a, the green pigment that is present in all algae. Klemas (2012) has illustrated the advantages and limitations of satellite and airborne remote sensing techniques for detecting phytoplankton and mapping algal blooms.

Choi et al. (2014) applied successfully the GOCI to monitor the distribution and temporal movement of the harmful algal bloom caused by *Cochlodinium polykrikoides* in the East Sea near the Korean Peninsula during August 2013. Details of the species have been described in Chap. 3 (Sect. 3.2.5). Harmful *C. polykrikoides* blooms have frequently appeared and caused fatal harm to aquaculture in Korean coastal waters since 1995. The authors investigated the applicability of GOCI, the world's first Geostationary Ocean Colour Imager, in monitoring the distribution and temporal movement of a harmful algal bloom (HAB) that was discovered in the East Sea near the Korean Peninsula in August 2013. They had identified the existence of *C. polykrikoides* at a maximum cell abundance of over 6000 cells/mL and a chlorophyll a concentration of over 400 mg/m^3. In areas of *C. polykrikoides* blooms, GOCI remote sensing reflectance (Rrs) spectra demonstrated the typical radiometric features of a HAB, and from the diurnal variations using GOCI-derived chlorophyll concentration images, we were able to identify the vertical migration of the red tide species. They had ascertained that the formation and propagation of the HAB had relations with cold water mass in the coastal region. GOCI can be effectively applied to the monitoring of short-term and long-term movements of red tides.

Satellite data can be used to effectively quantify cyanoHABs, and the highest abundances can be identified (Wynne et al. 2008) on a routine basis for multiple water bodies across transboundary regions. A previously validated algorithm was selected by Lunetta et al. (2015) with the objective of developing a robust temporal assessment method using MERIS (Medium Resolution Imaging Spectrometer). The data were used to quantify changes in cyanoHAB surface area extent in lakes in Florida, Ohio and California. The study was restricted to 4 years as continuous full-resolution MERIS imagery was only available between 2008 and 2012. The 2008–2012 MERIS archive may be the best historical record as a relative baseline for comparison against future events to determine change over time given the potential future availability of the operational Sentinel-3 OLCI sensors starting in 2017. Recently, Urquhart et al. (2017) developed a new assessment method using the programmable Medium Resolution Imaging Spectrometer (MERIS) imagery to quantify changes in inland cyanoHAB surface area across numerous water bodies in Florida, Ohio and California states. Results showed that in Florida, the spatial area of cyanoHABs increased largely mainly due to observed increases in high-risk bloom area.

1.6.1.6 Ocean Colour Monitor IRS-P4

Ocean Colour Monitor (OCM) of the Indian Remote Sensing Satellite IRS-P4 is optimally designed for multiple uses, such as (1) estimation of chlorophyll (phytopigment) in coastal and oceanic waters, (2) detection and monitoring of algal blooms, (3) studying the suspended sediment dynamics and (4) the characterization of the atmospheric aerosols. The technical specifications of the OCM sensor are given in Table 1.8. Sarangi and Gulshad Mohammed (2011) monitored the algal blooming in the chlorophyll images derived from the IRS-P4 OCM data around the southeast Arabian Sea and the movement mechanism of blooming features (algal filaments) during September 2002 and 2003. The authors have observed the red colour was extended of water up to 30–35 km from coastline and was reflected in OCM-derived chlorophyll images with the dense algal bloom features with very high chlorophyll concentration (20–50 mg/m^3).

So, the above satellite data sets have been useful in detecting and monitoring phytoplankton bloom in different oceans of the world. The blooms of diatoms, dinoflagellates, cyanophytes, coccolithophores and chlorophytes have been successfully detected and measured utilizing satellite remote sensing technology since many decades. Now, modelling the algal bloom based on in situ and satellite data sets has

Table 1.8 Technical characteristics of IRS-P4 OCM payload

Spectral range	**404–882 nm**
No. of channels	8
Wavelength range (nm)	Channel 1:404–423 (340.5)
Signal-to-noise ratio (SNR)	Channel 2:431–451 (440.7)
	Channel 3:475–495 (427.6)
	Channel 4:501–520 (408.8)
	Channel 5:547–565 (412.2)
	Channel 6:660–677 (345.6)
	Channel 7:749–787 (393.7)
	Channel 8:847–882 (253.6)
Satellite altitude (km)	720
Spatial resolution (m)	360 × 236
Swath (km)	1420
Repeativity	2 days
Quantization	12 bits
Equatorial crossing time	**12 noon**
Along track steering (to avoid sunglint)	20°

been the new challenge, which needs high-performance computing systems and large database of multiple parameters. Even of late researchers are trying to predict the algal bloom in different ecosystems of the World, based on several input parameters, hydrographic conditions like nutrient types, levels, light levels, turbidity, pigments, temperature, salinity etc. So, studying the phytoplankton bloom and categorizing it has been a tremendous challenge in modern biological oceanography in the advancement of technology and sophistication (Table 1.9).

1.7 Synoptic Account of Instruments Related to Algal Bloom Investigation

1.7.1 Liquid Chromatography Hyphenated with Tandem Mass Spectrometry (LC-MS/MS)

Liquid chromatography-mass spectrometry (LC-MS) is an analytical technique that coupled high-resolution chromatographic separation with sensitive and specific mass spectrum detection (Fig. 1.11). Typical LC-MS system is a combination of HPLC with MS using interface (ionization source) and is widely used in the field of bioanalysis due to their shorter analysis time, higher sensitivity and specificity. LC-MS is also used for the analysis of natural products and the profiling of secondary metabolites in plants. The sample is separated by LC, and the separated sample species are sprayed into atmospheric pressure ion source, where they are converted into ions in the gas phase. The mass analyser is then used to sort ions according to

Table 1.9 Detailed specifications of GOCI bands

Band (#)	Central wavelengths (nm)	Bandwidth (nm)	SNR	Primary application
Visible bands (1–6)				
1	412	20	1000	Yellow substance and turbidity
2	443	20	1090	Chlorophyll absorption maximum
3	490	20	1170	Chlorophyll and other pigments
4	555	20	1070	Turbidity, suspended sediment
5	660	20	1010	Atmospheric correction for turbid water, baseline of fluorescence signal, chlorophyll, suspended sediment
6	680	10	870	Fluorescence signal
Near-infrared bands (7–8)				
7	745	20	860	Atmospheric correction and baseline of fluorescence signal
8	865	40	750	Atmospheric correction, vegetation, water vapour reference over the ocean

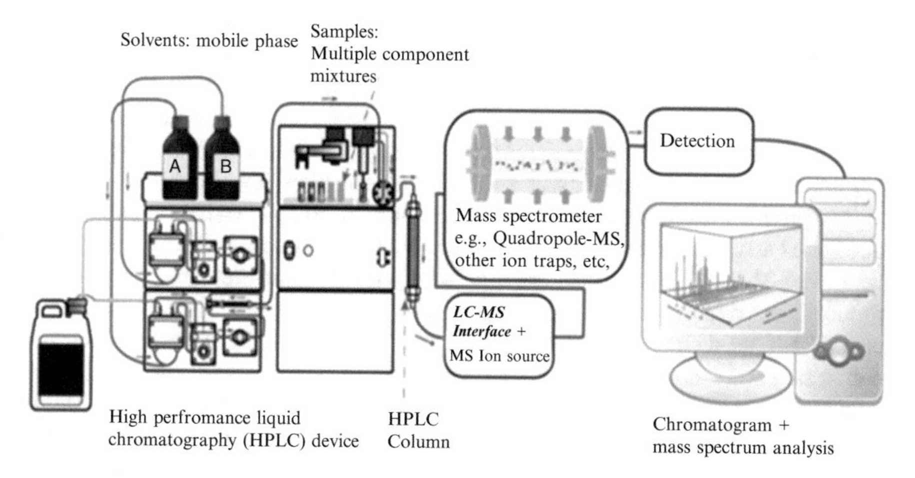

Fig. 1.11 Working principle of liquid chromatography hyphenated with tandem mass spectrometry (LC-MS/MS)

their mass-to-charge ratio, and detector counts the ions emerging from the mass analyser and may also amplify the signal generated from each ion. As a result, mass spectrum (a plot of the ion signal as a function of the mass-to-charge ratio) is created, which is used to determine the elemental or isotopic nature of a sample, the masses of particles and of molecules, and to elucidate the chemical structures of molecules.

It is frequently used in drug development as it allows quick molecular weight confirmation and structure identification. LC-MS applications for drug development are highly automated methods used for peptide mapping, glycoprotein mapping, natural product dereplication, bioaffinity screening, in vivo drug screening, metabolic stability screening, metabolite identification, impurity identification, quantitative bioanalysis and quality control. Recently, Procházková et al. (2017) developed and compared the solid-phase extraction and LC-MS/MS analytical approaches for the determination of phytoestrogens (eight flavonoids – biochanin A, coumestrol, daidzein, equol, formononetin, genistein, naringenin, apigenin – and five sterols, ergosterol, β-sitosterol, stigmasterol, campesterol, brassicasterol) and cholesterol in stagnant water bodies dominated by different cyanobacterial species.

Liquid chromatography hyphenated with mass spectrometry remains the best opinion for the quantitative determination of azaspiracids in the complex matrix of shellfish (Draisci et al. 2000; Quilliam et al. 2001). However, single-stage MS is prone to interferences, so a sample clean-up step is required. Ofuji et al. (1999) used solid-phase extraction (SPE) prior to the LC-MS determination of the marine toxins AZA1–AZA3. A detailed study of the clean-up efficiencies of a range of SPE phases was performed to complement LC-MS analysis of the azaspiracids (Moroney et al. 2002). Multiple tandem MS methods for the determination of azaspiracids have been developed using triple quadrupole instruments (Draisci et al. 2000; Volmer et al. 2002) and ion trap instruments (Furey et al. 2002; Lehane et al. 2002; Blay et al. 2003; James et al. 2003). Eleven azaspiracids were separated on a C18 column prior to MS^n identification (James et al. 2003; Lehane et al. 2004). However, chromatographic resolution is not necessary in LC-MS^n methods, if the structural differences between analogues are understood; AZA1–AZA10 were determined without the need for complete chromatographic separation. Using LC-MS^3 where the isomers were resolved using prescribed *mass* selection, it was possible to select unique productions for each azaspiracid (Lehane et al. 2002, 2004). Fux et al. (2007) developed an ultra-performance liquid chromatography (UPLC)-MS/MS method to determine 21 lipophilic marine toxins including AZA1, using preselected precursor-product ion *m/z* combinations; an analysis time of 6.6 min was reported. Rehmann et al. (2008) used UPLC combined with QqTOF MS/MS for the successful determination of known and new azaspiracid analogues in blue mussels; the group also employed an API 4000 QTrap system in their investigations (Rehmann et al. 2008).

Mouse and rat bioassays were the first methods used to determine AZA (Yasumoto et al. 1984). However, these methods were nonselective and lacked specificity (James et al. 2001). However, Hess et al. (2009) conducted a study to evaluate the ability of the mouse bioassay to detect the presence of AZA toxins at levels below the regulatory limit. Contaminated mussel hepatopancreas was mixed with toxin-free tissue to dilute the AZAs to appropriately low levels, and toxin concentrations were determined by LC-MS/MS; an intraday repeatability study was conducted on seven independent occasions over a 6-week period to evaluate the bioassay. The researchers found that the mouse bioassay delivered a high probability of producing a positive response around and below the regulatory limit. The test, though not quantitative, is adequate for the protection of human health, provided that the hepatopancreas is tested rather than total shellfish tissue (Hess et al. 2009).

1.7.2 *Optical Microscopy*

Optical microscopy is a technique employed to closely view a sample through the magnification of a lens with visible light. This is the traditional form of microscopy, which was first invented before the eighteenth century and is still in use today. An optical microscope, also recognized as a light microscope, uses one or a series of lenses to magnify images of small samples with visible light. The lenses are placed between the sample and the viewer's eye (eyepiece) to magnify the image so that it can be examined in greater detail. There are many types of optical microscopes. They can vary from a very basic design to a high complexity that offers higher resolution and contrast. The following types of optical microscopes are used for several purposes:

1. Simple microscope has a single lens to magnify the image of the sample, similar to a magnifying glass.
2. Compound microscope has a series of lenses to magnify the sample image to a higher resolution, more commonly used in modern research.
3. Digital microscope may have simple or compound lenses but uses a computer to visualize the image without the need for an eyepiece to view the sample.
4. Stereo microscope provides a stereoscopic image, which is useful for dissections.
5. Comparison microscope allows for the simultaneous view of two different samples, one in each eye.
6. Inverted microscope views the sample from underneath, which is useful to examine liquid cell cultures.

Other types of optical microscopes include petrographic, polarizing, phase-contrast, epifluorescence and confocal microscopes.

An optical microscope can generate a micrograph using standard light-sensitive cameras. Photographic film was traditionally used to capture the images. Technological developments have now enabled digital images to be taken with CMOS and charge-coupled device (CCD) cameras for optical microscopes. As a result, the image can be projected onto a computer screen in real time to examine a sample with these digital microscopes. This increases the convenience of use as eyepieces are no longer required. Optical microscopy is commonly used in diverse research fields including microbiology, microelectronics, nanophysics, biotechnology and pharmaceutical research. It can be also used to view biological samples for medical diagnoses, known as histopathology.

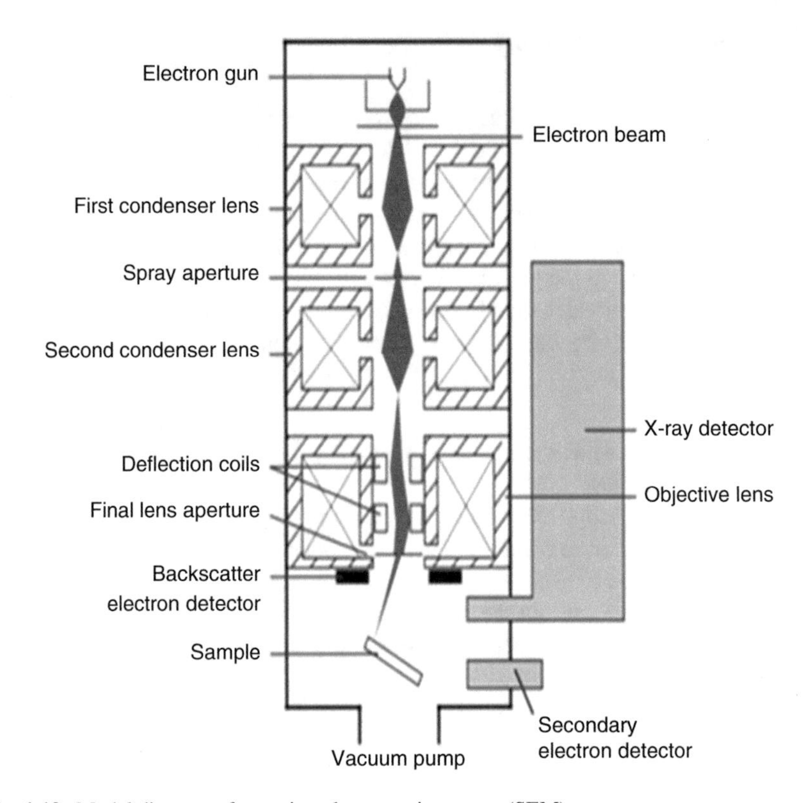

Fig. 1.12 Model diagram of scanning electron microscope (SEM)

1.7.3 Scanning Electron Microscope (SEM)

SEM is a powerful magnification tool that produces three-dimensional images with magnification of 10–200,000 times, hence much superior than optical microscope. It uses a beam of electron to form a large image of a very small object (e.g. microalgae). This can be used with thicker specimens and forms a perspective image.

A fine beam of electrons, focused by electromagnets, is moved or scanned across the specimen and works by scanning a focused beam of electrons on a sample of interest (Fig. 1.12). Electrons reflected from the specimen are collected by a detector, giving rise to an electrical signal, which is then used to generate a point of brightness on a TV-like screen. As the point moves regulating over the screen, in phase with the scanning electron beam, an image of the specimen is built up. Resolution power of the SEM is related on the electron beam size, and resolution is greater for finer beam.

The main SEM components include source of electrons, column down which electrons travel with electromagnetic lenses, electron detector, sample chamber and computer and display to view the images. Electrons are produced at the top of the column, accelerated down and passed through a combination of lenses and apertures to produce a focused beam of electrons which hits the surface of the sample. The sample is mounted on a stage in the chamber area, and unless the microscope is designed to operate at low vacuums, both the column and the chamber are evacuated by a combination of pumps. The level of the vacuum will depend on the design of the microscope. The position of the electron beam on the sample is controlled by scan coils situated above the objective lens. These coils allow the beam to be scanned over the surface of the sample. This beam rastering or scanning enables information about a defined area on the sample to be collected. As a result of the electron-sample interaction, a number of signals are produced. These signals are then detected by appropriate detectors. Magnification in a SEM can be controlled over a range of about 6 orders of magnitude from about 10 to 500,000 times.

SEM has the following diverse field of applications:

1. In biological research, SEMs can be used on anything from insect and animal tissue to bacteria and viruses. It includes measuring the effect of climate change of species, identifying new bacteria and virulent strains, testing vaccination, uncovering new species and working within the field of genetics.
2. In geological sampling, SEM can be used to determine weathering processes and morphology of the samples. Backscattered electron imaging can be used to identify compositional differences, while composition of elements can be provided by microanalysis. Valid uses include identification of tools and early human artefacts, soil quality measurement for farming and agriculture, dating historic ruins, forensic evidence in soil quality, toxins, etc.
3. In medical science, it is used to compare blood and tissue samples in determining the cause of illness and measuring the effects of treatments on patients (while contributing to the design of new treatments). Common uses include identifying diseases and viruses, testing new vaccinations and medicines and comparing tissue samples between patients in a control and test group.
4. Micrographs produced by SEMs have been used to create digital artworks. High-resolution three-dimensional images of various materials create a range of diverse landscapes and different images.

1.7.4 Nuclear Magnetic Resonance (NMR)

Nuclear magnetic resonance (NMR) spectroscopy, commonly known as NMR spectroscopy or *magnetic resonance* spectroscopy (MRS), is a powerful spectroscopy technique making better images of organs and soft tissue in comparison with other scanning techniques. This involves absorption of radiation at a photon energy, causing a transition from a lower to higher energy state. This research technique exploits the *magnetic* properties of certain atomic nuclei in which nuclei in a magnetic field absorb and re-emit electromagnetic radiation (as shown in Fig. 1.13). The main application of NMR is concerned to obtain physical, chemical, electronic and structural information about molecules, through selective absorption of very

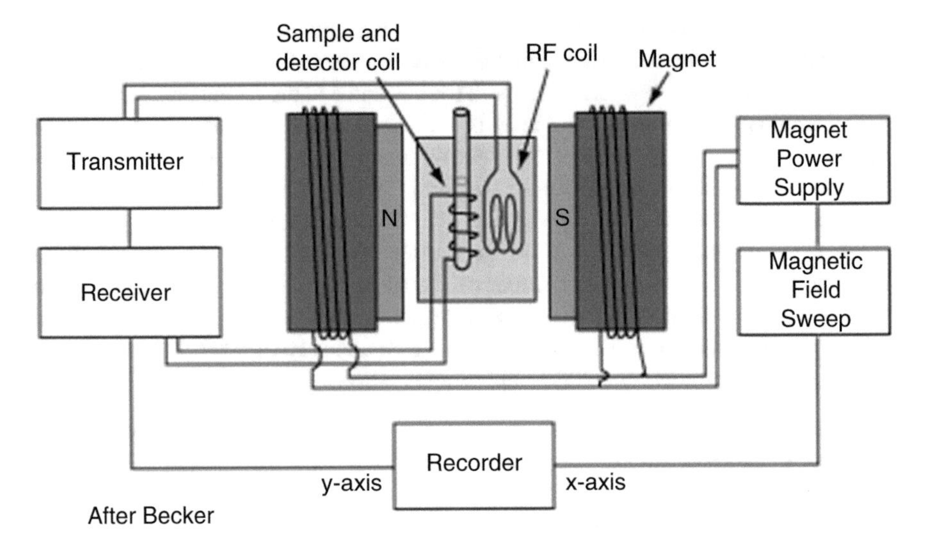

Fig. 1.13 Working principle of nuclear magnetic resonance (NMR) spectroscopy

high-frequency radio waves by certain atomic nuclei. Absorption of the radiation is detected when the difference between the nuclear levels corresponds to absorption of a quantum of radiation. An in-depth information on topology dynamics and three-dimensional structure of molecules both in solution and solid state can be done from NMR spectroscopy.

Is the absorption of radiation at a photon energy causing a transition from a lower to higher energy state? It is a physical phenomenon in which nuclei in a magnetic field absorb and re-emit electromagnetic radiation (Fig. 1.13). The theory behind the instrument comes from the spinning of the nucleus and generates a magnetic field. When an external magnetic field is present, the nuclei align themselves either with or against the field of the external magnet. If the external magnetic field is applied, an energy transfer occurs between the ground state and the excited state, and when the spin returns to its ground-state level, the absorbed radio-frequency energy is emitted at same frequency level. The emitted frequency signal gives the NMR spectrum of the concerned nucleus.

NMR spectroscopy is an analytical chemistry technique applied in quality control and for determining the content and purity of a sample as well as its molecular structure. It depends on the fact that the electrons in a molecule shield the nucleus to some extent from the field, causing different atoms to absorb at slightly different frequencies, recognized as chemical shifts. The main applications of NMR are as follows:

1. In biology, to study the biofluids, cells, perfused organs and biomacromolecules such as DNA, RNA, carbohydrates, proteins and peptides. It is used in the biochemistry laboratories for labelling studies. Moreover, Fourier transformation NMR is used for anatomical imaging, measuring physiological functions, flow measurements, angiography, etc.
2. In physics and physical chemistry, the high-pressure diffusion, liquid crystal solutions and rigid solids and membranes are studied.

3. In food science and pharmaceutical science, to study the drug metabolism and pharmaceuticals.
4. To elucidate chemical structures of organic and inorganic compounds and several ligand interactions.

Recently Poulin et al. (2018) studied the biological variability in allelopathic potency with the help of NMR spectroscopy to map the underlying chemical variation in the exuded metabolomes of five genetic strains of the dinoflagellate species, *Karenia brevis* (details of the species have been described in Chap. 3).

1.8 Methodology Used for Studying Algal Bloom

Next-generation sequencing (NGS), also known as high-throughput sequencing, is the catch-all term used to describe a number of different modern sequencing which allow us to sequence DNA and RNA much more quickly and cheaply for the study of genomics and molecular biology. Next-generation sequencing (NGS) is used for assessing phytoplankton diversity. Dinoflagellates are characterized by large genome size ranging from 1.5 to 185 Gbp. Genotyping-by-sequencing (GBS) method reduces genome complexity using restriction enzymes. Harmful microalgae genomes are described using various molecular approaches.

During the past decade, next-generation sequencing (NGS) technologies have provided new insights into the diversity, dynamics and metabolic pathways of natural microbial communities (bacteria and unicellular eukaryotes to microalgae). However, these new techniques face challenges related to the genome size and level of genome complexity of the species under investigation. Moreover, the coverage depth and the short-read length achieved by NGS-based approaches also represent a major challenge for assembly. These factors could limit the use of these sequencing methods for species lacking a reference genome and characterized by a high level of complexity.

Genotyping-by-Sequencing (GBS) method reduces genome complexity using restriction enzymes; in the context of genetic population studies, genotyping-by-sequencing (GBS), an NGS-based approach, could be used for the discovery and analysis of single nucleotide polymorphisms (SNPs). The NGS technologies are still relatively new and require further improvement. Specifically, there is a need to develop and standardize tools and approaches to handle large data sets, which have to be used for the majority of HAB species characterized by evolutionary highly dynamic genomes.

1.8.1 Mouse Bioassay

A mouse bioassay is a sensitive analytical method to determine biologically active toxins by observing the effects on living animals (in vivo) or in tissues (in vitro). This functional assay can either be qualitative (when the measured response is binary) or quantitative (when not binary), and this will provide simple information to confirm the presence of marine biotoxins (MB) standard test for laboratory confirmation of MBA botulinum toxin (BTX, a neurotoxic compound produced by *Clostridium botulinum*, a Gram-positive anaerobic bacterium) detection. MBA forms basis of most shellfish toxicity monitoring programme. Lab diagnosis of botulinum in horse traditionally has relied upon the MBA.

Recently, rapid, alternative, in vitro procedures have been developed for the detection of types A, B, E and F botulinal toxin-producing organisms and their toxins. The toxins generated in culture media can be detected using ELISA techniques such as the DIG-ELISA and the amp-ELISA. Biologically active and non-active toxins are detected since the assay detects the toxin antigen. The ELISA assays require 1 day of analysis. The toxin genes of viable organisms can be detected using the polymerase chain reaction technique and require 1 day of analysis after overnight incubation of botulinal spores or vegetative cells. In vitro assays that are positive are confirmed using the mouse bioassay.

The mouse assay was also used in a pilot study on PSP toxins in freeze-dried mussels, organized by the Food Analysis Performance Assessment Scheme (FAPAS®) in 2003 (Earnshaw 2003). Presently the mouse bioassay still forms the basis of most shellfish toxicity monitoring programme. The process was developed long before, subsequently modified and validated by the Association of Official Analytical Chemists (AOAC) for measurement of PSP toxins in a more rapid and accurate manner (Hollingworth and Wekell 1990).

The procedure was developed more than half a century ago and has been refined and standardized by the Association of Official Analytical Chemists (AOAC) to produce a rapid and reasonable accurate measurement of total PSP toxins (Hollingworth and Wekell 1990). Therefore major disadvantages of this assay are the lack of specificity (no differentiation between the various components of DSP toxins), subjectivity of death time of the animals and the maintaining and killing of laboratory animals. In addition, this assay is time-consuming and expensive, may give false positives because of interferences by other lipids (notably free fatty acids have shown to be very toxic to mice (Suzuki et al. 1996)) and shows variable results between whole body and hepatopancreas extracts (Botana et al. 1996; Van Egmond et al. 1993).

1.8.2 ELISA Test for PSP Toxin

The enzyme-linked immunosorbent assay, commonly known as ELISA or EIA, is a sensitive immunochemical technique that detects and measures a substance (antibodies) related to certain infectious conditions. This enzyme-based test can be of four different kinds: direct and indirect ELISA, sandwich ELISA and competitive ELISA. The test uses components of the immune system as well as chemicals for detection of immune responses in the body.

Paralytic shellfish poisoning is the most widespread algal-derived shellfish poisoning in the world. The toxins responsible for PSP are heterocyclic guanidines (saxitoxins); there are over 21 known congeners. PSP toxic syndrome is due primarily to the consumption of molluscan shellfish that have accumulated PSP toxins as a result of filter feeding on toxic dinoflagellates. Fish and crabs have also been implicated as vectors of PSP toxins. PSP toxin syndrome is characterized in its most severe form by paralysis of the breathing muscles, which if untreated could lead to death. At lower doses of the toxin, symptoms can range from mild stomach upset to a tingling sensation in the lips. The regulatory limit for PSP toxins is 40–80 μg PSP/100 grams of sample; these limits were established on bioassays measuring toxicity in mice. Mouse bioassays have some limitations, low sensitivity and variability of test results. There are some critics of the mouse toxicity method from animal protection groups. This ELISA test kit detects saxitoxins in water samples at the parts-per-trillion (ppt) levels, whereas saxitoxins on tissue samples can be detected at the ppb levels. Paralytic shellfish poisoning (PSP) toxins that act by blocking the sodium channel in neuronal and muscle cell membranes are a notorious cause of fatal food poisoning and are therefore under worldwide surveillance. Many attempts have been made to develop new methods to analyze PSP toxins and new toxicity-detecting systems which can take over the mouse bioassay and these are currently adopted as an official method worldwide. The cultured cell assay system, liquid chromatographic methods, radioimmunoassay and enzyme immunoassay have been developed and are known to be sensitive and time-saving compared with the mouse bioassay. However, most of these methods need expensive equipment, and some have a limited detection spectrum of the toxins. Direct enzyme-linked immunosorbent assay (ELISA) was introduced by Usleber et al. in 1994, and the system is now commercially available. The kit is relatively inexpensive and is convenient to use. The antibody used in this system was raised against saxitoxin (STX), which is the most potent PSP toxin, and the ELISA kit has a very high sensitivity to STX. Detection efficiency of an ELISA system for a toxin depends on the cross-reactivity of the antibody used in the system as the principle of ELISA is antigen-antibody binding.

1.9 Conclusion

Algal bloom abundance and fluctuations are commonly expressed in the context of algal biomass with chlorophyll amount. This can have large and varied negative impacts on marine ecosystems, depending on the species involved (both innocuous and harmful), the geographical locations where they are found and the mechanism by which they exert such effects. The event causes considerable ecological disturbance, economic losses and public health concern. Most importantly, when essential nutrients are exhausted by the algae and the algal biomass decays, the bacterial decomposition of large amounts of organic material depletes the available oxygen, and fish may die as a result of hypoxic conditions.

It is evident that traditional monitoring process can be substantiated by satellite imagery which is a modern practical tool for spatiotemporal coverage of the surface layer (photic layer) of the water column. Remote sensing proves a good way to allow assessment of environmental change over long periods of time and to help understand the temporal patterns of chlorophyll in response to physical fluctuations. There is substantial improvement in remote sensing technologies, being successful for frequent synoptic water quality monitoring process over large areas and representing the most convenient option to measure the phytoplankton community to nutrient enrichment.

References

Anantharaman, P., Thirumaran, G., Arumugam, R., Ragupathi Raja Kannan, R., Hemalatha, A., Kannathasan, A., Sampathkumar, P., & Balasubramanian, T. (2010). Monitoring of Noctilucabloom in Mandapam and Keelakarai coastal waters; South-East coast of India. *Recent Research in Science and Technology, 2*(10), 51–58.

Anderson, D. M., Glibert, P. M., & Burkholder, J. M. (2002). Harmful algal blooms and eutrophication: Nutrient sources, composition and consequences. *Estuaries, 25*(4B), 704–726.

Anoop, A. K., Krishnakumar, P. K., & Rajagopalan, M. (2007). Trichodesmium erythraeum (Ehrenberg) bloom along the southwest coast of India (Arabian Sea) and its impact on trace metal concentrations in seawater. *Estuarine, Coast and Shelf Science, 71*, 641–646.

Aspila, K. I., Agemian, T. I., & Chan, A. S. (1976). Semi-automatic method for the determination of inorganic, organic and total phosphorous in sediments. *Analyst, 101*, 187–197.

Bastviken, D., & Tranvik, L. (2001). The leucine incorporation method estimates bacterial growth equally well in both oxic and anoxic lake waters. *Applied and Environmental Microbiology, 67*, 2916–2921.

Blay, P. K. S., Brombacher, S., & Volmer, D. A. (2003). Studies on azaspiracidbiotoxins. III. Instrumental validation for rapid quantification of AZA 1 in complex biological matrices. *Rapid Communications in Mass Spectrometry, 17*, 2153–2159.

Botana, L. M., Rodriguez-Vieytes, M., Alfonso, A., & Louzao, M. C. (1996). Phycotoxins: Paralytic shellfish poisoning and diarrheic shellfish poisoning. In L. M. L. Nollet (Ed.), *Handbook of food analysis – Residues and other food component analysis* (Vol. 2, pp. 1147–1169). New York: Marcel Dekker Inc.

Carder, K. L., Chen, F. R., Lee, Z. P., Hawes, S. K., & Kamykowski, D. (1999). Semianalytic moderate-resolution imaging spectrometer algorithms for chlorophyll a and absorption with bio-optical domains based on nitrate-depletion temperatures. *Journal of Geophysical Research, 104*, 5403–5421.

Carpenter, S. R., Caraco, N. F., Correll, D. L., Howarth, R. W., Sharpley, A. N., & Smith, V. H. (1998). Nonpoint pollution of surface waters with phosphorus and nitrogen. *Ecological Applications, 8*(3), 559–568.

Casabianca, S., Cornetti, L., Capellacci, S., Vernesi, C., & Penna, A. (2017). Genome complexity of harmful microalgae. *Harmful Algae, 63*, 7–12.

Choi, J. K., Min, J. E., Noh, J. H., Han, T. H., Yoon, S., Park, Y. J., & Park, J. H. (2014). Harmful algal bloom (HAB) in the East Sea identified by the Geostationary Ocean Color Imager (GOCI). *Harmful Algae, 39*, 295–302.

Cloern, J. E. (2001). Our evolving conceptual model of the coastal eutrophication problem. *Marine Ecology Progress Series, 210*, 223–253.

Conley, D. J. (1999). Biogeochemical nutrient cycles and nutrient management strategies. In *Man and river systems* (pp. 87–96). Dordrecht: Springer.

Dale, B. (1976). Cyst formation, sedimentation, and preservation: Factors affecting dinoflagellate assemblages in recent sediments from Trondheims fjord, Norway. *Review of Palaeobotany and Palynology, 22*, 39–60.

Dale, B. (1983). Dinoflagellate resting cysts: 'Benthic plankton'. In G. A. Fryxell (Ed.), *Survival strategies of the algae* (pp. 69–137). Cambridge: Cambridge University Press.
Dale, B. (1996). Dinoflagellate cyst ecology: Modeling and geological applications. In J. Jansonius & D. C. McGregor (Eds.), *Palynology principles and applications* (pp. 1249–1275). Dallas: American Association of Stratigraphic Palynologists Foundation.
Dale, B. (2009). Eutrophication signals in the sedimentary record of dinoflagellate cysts in coastal waters. *Journal of Sea Research, 61*, 103–113.
De Vernal, A., Rochon, A., Turon, J. L., & Matthiessen, J. (1998). Organic-walled dinoflagellate cysts: Palynological tracers of sea-surface conditions in middle to high latitude marine environments. *Geobios, 30*, 905–920.
Devassy, V. P. (1974). Observation on the bloom of a diatom FragilariaoceanicaCleve. *Mahasagar, 7*, 101–105.
Devassy, V. P., & Bhattathiri, P. M. A. (1974). Phytoplankton ecology of the Cochin backwaters. *Indian Journal of Marine Science, 3*, 46–50.
Devassy, V. P., & Sreekumaran Nair, S. R. (1987). Discolouration of water and its effect on fisheries along the Goa coast. *Mahasagar-Bulletin of the National Institute of Oceanography, 20*(2), 121–128.
Diaz, R. J., & Rosenberg, R. (2008). Spreading dead zones and consequences for marine ecosystems. *Science, 321*(5891), 926–929.
Domingues, R. B., Barbosa, A., & Galvão, H. (2008). Constraints on the use of phytoplankton as a biological quality element within the Water Framework Directive in Portuguese waters. *Marine Pollution Bulletin, 56*, 1389–1395.
Dortch, Q., & Whitledge, T. E. (1992). Does nitrogen or silicon limit phytoplankton production in the Mississippi River plume and nearby regions? *Continental Shelf Research, 12*, 1293–1309.
Draisci, R., Palleschi, L., Ferretti, E., Furey, A., James, K. J., Satake, M., & Yasumoto, T. (2000). Development of a liquid chromatography– Tandem mass spectrometry method for the identification of azaspir acid in shellfish. *Journal of Chromatography, 871*, 13–21.
Earnshaw, A. (2003). *Marine toxins, Pilot Study August 2003* (Report food analysis performance assessment scheme). Sand Hutton: Central Science Laboratory.
Engström-Öst, J., Brutemark, A., Vehmaa, A., Motwani, N. H., & Katajisto, T. (2015). Consequences of a cyanobacteria bloom for copepod reproduction, mortality and sex ratio. *Journal of Plankton Research, 37*(2), 388–398.
EL Wakeel, S. K., & Riley, J. P. (1957). Determination of organic carbon in marine mud. *Journal du Conseil International pour Exploration de la Mer, 22*, 180–183.
Fenchel, T., Kristensen, L. D., & Rasmussen, L. (1990). Water column anoxia – Vertical zonation of planktonic protozoa. *Marine Ecology Progress Series, 62*, 1–10.
Fensome, R.A., Taylor, F.J.R., Norris, G., Sarjeant, W.A.S., Wharton, D.I., Williams, G.L., 1993. *A classification of living and fossil dinoflagellates* (Special Publication Number 7, pp. 351). New York: Micropaleontology Press, American Museum of Natural History.
Folk, R. L., & Ward, W. C. (1957). Brazon river bars, a study in significance of grain-size parameters. *Journal of Sedimentary Petrology, 27*, 3–27.
Furey, A., Braña-Magdalena, A., Lehane, M., Moroney, C., James, K. J., Satake, M., & Yasumoto, T. (2002). Determination of azaspiracids in shellfish using liquid chromatography/tandem electrospray mass spectrometry. *Rapid Communications in Mass Spectrometry, 16*(3), 238–242.
Fux, E., McMillan, D., Bire, R., & Hess, P. (2007). Development of an ultra- performance liquid chromatography–mass spectrometry method for the detection of lipophilic marine toxins. *Journal of Chromatography A, 1157*, 273–280.
Gohin, F., Saulquin, B., Oger-Jeanneret, H., Lozac'h, L., Lampert, L., Lefebvre, A., & Bruchon, F. (2008). Towards a better assessment of the ecological status of coastal waters using satellite-derived chlorophyll-a concentrations. *Remote Sensing of Environment, 112*(8), 3329–3340.
Gopakumar, G., Sulochanan, B., & Venkatesan, V. (2009). Bloom of Noctilucascintillans (Maccartney) in Gulf of Mannar, southeast coast of India. *Journal of Marine Biological Association of India, 55*(1), 75–80.
Hallegraeff, G. M. (1993). A review of harmful algal blooms and their apparent global increase. *Phycologia, 32*(2), 79–99.

Hansen, J. L. S., & Bendtsen, J. (2009). Effects of climate change on hypoxia in the North Sea – Baltic Sea transition zone. *Earth and Environmental Science, 6*, 1755–1307.
Head, M. J. (1996). Modern dinoflagellate cysts and their biological affinities. In J. Jansonius & D. C. McGregor (Eds.), *Palynology: Principles and applications* (Vol. 3, pp. 1197–1248). Salt Lake City: A.A.S.P. Foundation.
Hess, P., Butter, T., Petersen, A., Silke, J., & McMahon, T. (2009). Performance of the EU-harmonised mouse bioassay for lipophilic toxins for the detection of azaspiracids in naturally contaminated mussel (*Mytilus edulis*) hepatopancreas tissue homogenates characterized by liquid chromatography coupled to tandem mass spectrometry. *Toxicon, 53*(7), 713–722.
Hollingworth, T., & Wekell, M. M. (1990). Fish and other marine products 959.08. Paralytic shellfish poisoning. Biological method, final action. In K. Hellrich (Ed.), *Official methods of analysis of the Association of Official Analytical Chemists* (15th ed., pp. 881–882). Arlington: Association of Official Analytical Chemists.
Hornell, J., & Nayudu, R. M. (1923). A contribution to the life history of sardines with notes on the plankton of the Malabar Coast. *Madras Fisheries Bulletin, 17*, 129–197.
Howarth, R. W., Anderson, D., Cloern, J., Elfring, C., Hopkinson, C., Lapointe, B., Malone, T., Marcus, N., McGlathery, K., Sharpley, A., & Walker, D. (2000). Nutrient pollution of coastal rivers, bays, and seas. *Issues. Ecology, 7*, 1–15.
ICES. (1984). *Report of the ICES special meeting on the causes, dynamics and effects of exceptional marine blooms and related events* (International Council Meeting Paper 1984/ E, 42, ICES.
IOCCG. (2000). Remote sensing of ocean color in coastal, and other optically complex waters. In S. Sathyendranath (Ed.), *Reports of the International Color Coordinating Group 3* (p. 144). Dartmouth: IOCCG Project Office.
Ismael, A. A. (1993). *Systematic and ecological studies of the planktonic dinoflagellates of the coastal water of Alexandria*. M.Sc. Thesis, Faculty of Science in Alexandria University, p. 115.
Ismael, A. A., & Khadr, A. (2003). *Alexandrium minutum* cysts in sediment cores from the Eastern Harbour of Alexandria, Egypt. *Oceanologia, 45*(4), 721–731.
Ismael, A., El-Masry, E., & Khadr, A. (2014). Dinoflagellate [cyst] [S1] as signals for eutrophication in the eastern harbour of Alexandria-Egypt. *Indian Journal of Geo-Marine Sciences, 43*(3), 365–371.
Iyer, C. S. P., Robin, R. S., Sreekala, M. S., & Kumar, S. S. (2008). Kareniamikimotoi bloom in Arabian Sea. *Harmful Algae News, 37*, 9–10.
James, K. J., Furey, A., Lehane, M., Moroney, C., Satake, M., & Yasumoto, T. (2001). LC-MS methods for the investigation of a new shellfish toxic syndrome–Azaspiracid Poisoning (AZP). In *Mycotoxins and phycotoxins in perspective at the turn of the century* (pp. 401–408). Wageningen.
James, K. J., Sierra, M. D., Lehane, M., Magdalena, A. B., & Furey, A. (2003). Detection of five new hydroxyl analogues of azaspiracids in shellfish using multiple tandem mass spectrometry. *Toxicon, 41*, 277–283.
Justic, D., Rabalais, N. N., & Turner, R. E. (1995). Stoichiometric nutrient balance and origin of coastal eutrophication. *Marine Pollution Bulletin, 30*(1), 41–46.
Kanaya, G, Nakamura, Y, & Koizumi, T. (2018). Ecological thresholds of hypoxia and sedimentary H_2S in coastal soft-bottom habitats: A macro invertebrate-based assessment. *Marine Environmental Research* (article in press).
Karunasagar, I. (1993). Gymnodinium kills farm fish in India. *Harmful Algae News, 5*(3).
Karunasagar, I., Gowda, H. S. V., Subburaj, M., Venugopal, M. N., & Karunasagar, I. (1984). Outbreak of paralytic shellfish poisoning in Mangalore, west coast of India. *Current Science, 53*, 247–249.
Katti, R. J., Gupta, T. R. C., & Shetty, H. P. C. (1988). On the occurrence of "green tide" in the Arabian Sea off Mangalore. *Current Science, 57*, 38–381.
Klemas, V. (2012). Remote sensing of algal blooms: An overview with case studies. *Journal of Coastal Research, 28*(1A), 34–43.
Krumbein, W. C., & Pettijohn, F. J. (1938). *Manual of sedimentary petrology* (p. 549). New York: Appleton Century and Crofts.

Lee, Z., Carder, K. L., Florida, S., & Petersburg, S. (2005). Hyperspectral remote sensing. In *Remote sensing of coastal aquatic environments technologies, techniques and applications* (pp. 181–204). Dordrecht: Springer.
Lehane, M., Brana-Magdalena, A., Moroney, C., Furey, A., & James, K. J. (2002). Liquid chromatography with electrospray ion trap mass spectrometry for the determination of five azaspiracids in shellfish. *Journal of Chromatography A, 950*, 139–147.
Lehane, M., Saez, M. J. F., Magdalena, A. B., Canas, I. R., Sierra, M. D., Hamilton, B., Furey, A., & James, K. J. (2004). Liquid chromatography– Multiple tandem mass spectrometry for the determination of ten azaspiracids, including hydroxyl analogues in shellfish. *Journal of Chromatography A, 1024*, 63–70.
Lucas, C. E. (1947). The ecological effects of external metabolites. *Biological Reviews, 22*(3), 270–295.
Lunetta, R., Schaeffer, B., Stumpf, R., Keith, D., Jacobs, S., & Murphy, M. (2015). Evaluation of cyanobacteria cell count detection derived from MERIS imagery across the eastern USA. *Remote Sensing of Environment, Elsevier, 157*, 24–34.
Madhu, N. V., Reny, P. D., Paul, M., Ulhas, N., & Resmi, P. (2011). Occurrence of red tide caused by Kareniamikimotoi (toxic dinoflagellate) in the south-west coast of india. *Indian Journal of Geo-Marine Sciences, 40*(6), 821–825.
Marret, F., & Zonneveld, K. A. F. (2003). Atlas of modern organic- walled dinoflagellate cyst distribution. *Review of Palaeobotany and Palynology, 125*, 1251–1200.
Matsuoka, K. (1985). Sustained oscillations generated by mutually inhibiting neurons with adaptation. *Biological Cybernetics, 52*, 367–376.
Matsuoka, K. (1987). Mechanisms of frequency and pattern control in the neural rhythm generators. *Biological Cybernetics, 56*, 345–353.
Matsuoka, K. (1999). Eutrophication process recorded in dinoflagellate cyst assemblages a case of Yokohama Port, Tokyo Bay. *Japan Science of the Total Environment, 231*, 17–35.
Matsuoka, K. (2001). Further evidence for a marine dinoflagellate cyst as an indicator of eutrophication process in Yokohama Port, Tokyo Bay, Japan. Comments on a discussion by B. Dale. *Japan Science of the Total Environment, 264*, 221–233.
Matsuoka, K., & Fukuyo, Y. (2000). *Technical guide for modern Dinoflagellate cyst study* (p. 29). WESTPAC- HAB/WESTPAC/IOC, Japan Society for the Promotion of Science.
Matsuoka, K., Joyce, L. B., Kotani, Y., & Matsuyama, Y. (2003). Modern dinoflagellate cysts in hypertrophic coastal waters of Tokyo Bay, Japan. *Journal of Plankton Research, 25*, 1461–1470.
Meire, L., Soetaert, K. E. R., & Meysman, F. J. R. (2013). Impact of global change on coastal oxygen dynamics and risk of hypoxia. *Biogeosciences, 10*(4), 2633–2653.
Mohanty, A. K., Satpathy, K. K., Sahu, G., Sasmal, S. K., Sahu, B. K., & Panigrahy, R. C. (2007). Red tide of Noctilucascintillans and its impact on the coastal water quality of the near-shore waters, off the Rushikulya River, Bay of Bengal. *Current Science, 93*, 616–618.
Mora, C., et al. (2013). Biotic and human vulnerability to projected changes in ocean biogeochemistry over the 21st century. *PLoS Biology, 11*, e1001682.
Morel, A., & Prieur, L. (1977). Analysis of variations in ocean color. *Limnology and Oceanography, 22*, 709–722.
Moroney, C., Lehane, M., Brana-Magdalena, A., Furey, A., & James, K. J. (2002). Comparison of solid-phase extraction methods for the determination of azaspiracids in shellfish by liquid chromatography–electrospray mass spectrometry. *Journal of Chromatography A, 963*, 353–361.
Mudroch, A., Azcue, J. M., & Mudroch, P. (1997). *Influence of the use of a drainage basin on physical and chemical properties of bottom sediments of lakes* (p. 287). Boca Raton: Lewis Publishers, CRC Press, Inc.
Murata, A., Nagashima, Y., & Taguchi, S. (2012). N: P ratios controlling the growth of the marine dinoflagellate *Alexandrium tamarense*: Content and composition of paralytic shellfish poison. *Harmful Algae, 20*, 11–18.
Nagvi, S. W. A., Yoshinari, T., Jayakumar, D. A., Altabet, M. A., Narvekar, R. V., Devol, A. H., Brandes, J. A., & Codispoti, L. A. (1998). Budgetary and biogeochemical implications of N2O isotope signatures in the Arabian Sea. *Nature, 394*, 462–464.

Nayak, B. B., Karunasagar, I., & Karunasagar, I. (2000). Bacteriological and physico-chemical factors associated with Noctilucamilliaris bloom, along Mangalore, southwest coast of India. *Indian Journal of Marine Science, 29*(2), 139–143.

Nixon, S. W. (1995). Coastal marine eutrophication – A definition, social causes, and future concerns. *Ophelia, 41*, 199–219.

Nixon, S. W. (1996). Regional coastal research—What is it? Why do it? What role should NAML play? *Biological Bulletin, 190*, 252–259.

Novoa, S., Chust, G., Valencia, V., Froidefond, J.-M., & Morichon, D. (2011). Estimation of chlorophyll-a concentration in waters over the continental shelf of the Bay of Biscay: A comparison of remote sensing algorithms. *International Journal of Remote Sensing, 32*, 8349–8371.

O'Herald. (2001). NIO discovers toxic algal off Goa. *O'Herald Newspaper Goa*.

O'Neil, J. M. (2012). The psychology of men. In E. Altmaier & J. Hansen (Eds.), *Oxford handbook of counseling psychology*. New York: Oxford University Press.

Ofuji, K., Satake, M., Oshima, Y., McMahon, T., James, K. J., & Yasumoto, T. (1999). A sensitive and specific determination method for azaspiracids by liquid chromatography mass spectrometry. *Natural Toxins, 7*, 247–250.

Oliver, M. (2017). Meat industry blamed for largest-ever 'dead zone' in Gulf of Mexico. *The Guardian*. ISSN 0261-3077.

Padmakumar, K. B., Sanilkumar, M. G., Saramma, A. V., Sanjeevan, V. N., & Menon, N. R. (2007). Green tide of Noctilucamiliarisin the Northern Arabian Sea. *Harmful Algae News*, 1–16.

Padmakumar, K. B., Sanilkumar, M. G., Saramma, A. V., Sanjeevan, V. N., & Menon, N. R. (2008a). Microcystisaeruginosa bloom on Southwest coast of India. *Harmful Algae News, 37*, 11–12.

Padmakumar, K. B., Sanilkumar, M. G., Saramma, A. V., Sanjeevan, V. N., & Menon, N. R. (2008b). "Green tide" of Noctilucamiliaris in the Northern Arabian Sea. *Harmful Algae News, 36*, 12.

Padmakumar, K. B., Smitha, B. R., Thomas, L. C., Fanimol, C. L., SreeRenjima, G., Menon, N. R., & Sanjeevan, V. N. (2010). Blooms of Trichodesmium erythraeum in the South Eastern Arabian Sea during the onset of 2009 summer monsoon. *Ocean Science Journal, 45*(3), 151–157.

Padmakumar, K. B., Thomas, L. C., Salini, T. C., John, E., Menon, N. R., & Sanjeevan, V. N. (2011). Monospecific bloom ofnoxious raphidophyteChattonella marina in the coastal waters of South-West coast of India. *International Journal of Biosciences, 1*(1), 57–69.

Pearl, H. W., & Huisman, J. (2009). Climate change: A catalyst for global expansion of harmful cyanobacterial blooms. *Environmental Microbiology, 1*, 27–37.

Pearl, H. W., & Otten, T. G. (2013). Harmful cyanobacterial blooms: Causes, consequences, and controls. *Microbial Ecology, 65*, 995–1010.

Pickard, G. L., & Emery, W. J. (1982). *Description physical oceanography: An introduction* (p. 47). Oxford: Pergamon Press.

Pospelova, V., & Chmura, G. L. (1998). Modern distribution of dinoflagellate cysts in coastal lagoons of Rhode Island, USA. *Norges teknisk-naturvitenskapelige universitet Vitenskapsmuseet Rapport Botanisk Serie, 1*, 122–123.

Pospelova, V., & Head, M. J. (2002). *Islandinium brevispinosum* sp. nov. (Dinoflagellata), a new species of organic-walled dinoflagellate cyst from modern estuarine sediments of New England (USA). *Journal of Phycology, 38*, 593–601.

Pospelova, V., & Kim, S. J. (2010). Dinoflagellate cysts in recent estuarine sediments from aquaculture sites of southern South Korea. *Marine Micropaleontology, 76*, 37–51.

Poulin, R. X., Poulson-Ellestad, K. L., Roy, J. S., & Kubanek, J. (2018). Variable allelopathy among phytoplankton reflected in red tide metabolome. *Harmful Algae, 71*, 50–56.

PrabhuMatondkar, S. G., Bhat, S. R., Dwivedi, R. M., & Nayak, S. R. (2004). Indian satellite IRS-P4 (OCEANSAT). Monitoring algal blooms in the Arabian Sea. *Harmful Algae News, 26*, 4–5.

Prakash, A., & Sarma, A. H. V. (1964). On the occurrence of "red water"phenomenon on the west coast of India. *Current Science, 33*, 168–170.

Procházková, T., Sychrová, E., Javůrková, B., Večerková, J., Kohoutek, J., Lepšová-Skácelová, O., & Hilscherová, K. (2017). Phytoestrogens and sterols in waters with cyanobacterial blooms-analytical methods and estrogenic potencies. *Chemosphere, 170*, 104–112.

Quilliam, M. A., Hess, P., & Dell'Aversano, C. (2001). Recent developments in the analysis of phycotoxins by liquid chromatography–mass spectrometry. In *Mycotoxins and phycotoxins in perspective at the turn of the century* (pp. 383–391). Wageningen: Ponsen and Looijen.

Rajasekar, K. T., Rajkumar, M., Sun, J., Prabu, V. A., & Perumal, P. (2010). Bloom forming species of phytoplankton in two coastal waters in the Southeast coast of India. *Journal of Ocean University of China, 9*(3), 265–272.

Redfield, A. C. (1934). In R. J. Daniel (Ed.), *James Johnstone memorial volume*. Liverpool: Liverpool University Press.

Redfield, A. C. (1958). The biological control of chemical factors in the environment. *American Scientist, 46*, 205–221.

Rehmann, N., Hess, P., & Quilliam, M. A. (2008). Discovery of new analogs of the marine biotoxin azaspiracid in blue mussels (*Mytilus edulis*) by ultra-performance liquid chromatography/tandem mass spectrometry. *Rapid Communications in Mass Spectrometry, 22*(4), 549–558.

Ryther, J. H., & Dunxtan, W. M. (1971). Nitrogen, phosphorus, and eutrophication in the coastal marine environment. *Science, 171*(3975), 1008–1013.

Ryu, K., Lee, H.-R., & Kim, W. G. (2012). The influence of the quality of the physical environment, food, and service on restaurant image, customer perceived value, customer satisfaction, and behavioral intentions. *International Journal of Contemporary Hospitality Management, 24*(2), 200–223.

Saetre, M. M. L., Dale, B., Abdullah, M. I., & Saetre, G. P. O. (1997). Dinoflagellate cysts as potential indicators of industrial pollution in a Norwegian fjord. *Marine Environmental Research, 44*(2), 167–189.

Sahayak, S., Jyothibabu, R., Jayalakshmi, K. J., Habeebrehman, H., Sabu, P., Prabhakaran, M. P., Jasmine, P., Shaiju, P., George, R. M., Threslamma, J., & Nair, K. K. C. (2005). Red tide of Noctilucamiliaris off south of Thiruvananthapuram subsequent to the 'stench event' at the southern Kerala coast. *Current Science, 89*, 1472–1473.

Sanilkumar, M. G., Thomas, A. M., Philip, A. A., Hatha, M., Sanjeevan, V. N., & Saramma, A. V. (2009). First report of Protoperidinium bloom from Indian waters. *Harmful Algae News, 39*, 15.

Santhosh Kumar, C., Ashok Prabu, V., Sampathkumar, P., & Anantharaman, P. (2010). Occurrence of algal bloom Microcystisaeruginosa in the Vellar estuary, South-East coast of India. *International Journal of Current Research, 5*, 52–55.

Sarangi, R. K., & Mohamed, G. (2011). Seasonal algal bloom and water quality around the coastal Kerala during southwest monsoon using in situ and satellite data. *Indian Journal of Geo-Marine Sciences, 40*(3), 356–369.

Sahu, S. K., & Mukhopadhyay, S. (2015). *On generating a flexible class of anisotropic spatial models using gaussian predictive processes*. Technical Report, University of Southampton.

Satpathy, K. K., Mohanty, A. K., Sahu, G., Prasad, M. V. R., Venkatesan, R., Natesan, U., & Rajan, M. (2007). On the occurrence of Trichodesmium erythraeum (Ehr.) bloom in the coastal waters of Kalpakkam, east coast of India. *Indian Journal of Science and Technology, 1*(2), 1–9.

Schindler, D. W. (1974). Eutrophication and recovery in experimental lakes: Implications for lake management. *Science, 184*(4139), 897–899.

Schindler, D. W. (2006). Recent advances in the understanding and management of eutrophication. *Limnology and Oceanography, 51*(1part2), 356–363.

Smayda, T. J. (1989). Primary production and the global epidemic of phytoplankton blooms in the sea: A linkage? In E. M. Cosper, V. M. Bricelj, & E. J. Carpenter (Eds.), *Novel phytoplankton blooms* (pp. 449–483). Berlin: Springer.

Smayda, T. J. (1990). Novel and nuisance phytoplankton blooms in the sea: Evidence for a global epidemic. In E. Graneli, B. Sundstrom, L. Edler, & D. M. Anderson (Eds.), *Toxic marine phytoplankton* (pp. 29–40). New York: Elsevier.

Smayda, T. J. (2004). Eutrophication and phytoplankton. In P. Wassmann, & K. Olli (Eds.), *Integrated approaches to drainage basin nutrient inputs and coastal eutrophication* (pp. 89–98). Electronic-Book (pdf file on web page: http://www.ut.ee/_olli/eutr/).

Smayda, T. J., & White, A. W. (1990). Has there been a global expansion of algal blooms? If so is there a connection with human activities? In E. Granelli (Ed.), *Toxic marine phytoplankton* (p. 516). New York: Elsevier.

Stolte, W., McCollin, T., Noordeloos, A. A. M., & Riegman, R. (1994). Effect of nitrogen source on the size distribution within marine phytoplankton populations. *Journal of Experimental Marine Biology and Ecology, 184*, 83–97.

Suikkanen, S., Pulina, S., Engström-Öst, J., Lehtiniemi, M., Lehtinen, S., & Brutemark, A. (2013). Climate change and eutrophication induced shifts in northern summer plankton communities. *PLoS One, 8*(6), 466–475.

Suzuki, T. M., Sherr, B. E., & Sherr, F. B. (1996). Estimation of ammonium regeneration efficiencies associated with bacterivory in pelaglc food webs via a 15N tracer method. *Journal of Plankton Research, 18*, 411–428.

Syrett, P. J. (1981). Nitrogen metabolism of microalgae. *Canadian Bulletin of Fisheries and Aquatic Sciences, 210*, 182–210.

Taylor F. J. R., Pallingher U., 1987. Ecology of dinoflagellates. In: F. J. R. Taylor (Ed.), *The biology of dinoflagellates* (Botanical monographs, Vol. 21, pp. 399–529). Oxford: Blackwell Scientific.

Thorsen, T. A., & Dale, B. (1997). Dinoflagellate cysts as indicators of pollution and past climate in a Norwegian fjord. *Holocene, 7*(4), 433–446.

Tiwari, L. R., & Nair, V. R. (1998). Ecology of phytoplankton from Dharmatar creek, west coast of India. *Indian Journal of Marine Science, 27*(3&4), 302–309.

Turner, R. E., Qureshi, N., Rabalais, N. N., Dortch, Q., Justic, D., Shaw, R., & Cope, J. (1998). Fluctuating silicate: Nitrate ratios and coastal plankton food webs. *Proceedings of the National Academy of Sciences of the United States of America, 95*, 13048–13051.

Turner, M. G., Gardner, R. H., & O'Neill, R. V. (2001). *Landscape ecology in theory and practice* (p. 401). New York: Springer.

Urquhart, E. A., Schaeffer, B. A., Stumpf, R. P., Loftin, K. A., & Jeremy Werdell, P. (2017). A method for examining temporal changes in cyanobacterial harmful algal bloom spatial extent using satellite remote sensing. *Harmful Algae, 67*, 144–152.

Usleber, E., Straka, M., & Terplan, G. (1994). Enzyme immunoassay for fumonisin B1 applied to cornbased food. *Journal of Agricultural and Food Chemistry, 42*, 1392–1396.

Van Dolah, F. M. (2000). Marine algal toxins: Origins, health effects, and their increased occurrence. *Environmental Health Perspectives, 108*(1), 133–141.

Van Egmond, H. P., Aune, T., Lassus, P., Speijers, G., & Waldock, M. (1993). Paralytic and diarrheic shellfish poisons: Occurrence in Europe, toxicity, analysis and regulation. *Journal of Natural Toxins, 2*, 41–83.

Vaquer-Sunyer, R., & Duarte, C. M. (2008). Threshholds of hypoxia for marine biodiversity. *Proceedings of the National Academy of Sciences, 105*(40), 15452–15457.

Venugopal, P., Haridas, P., MadhuPratap, M., SRao, S., & T. (1979). Incidence of red water along south Kerala coast. *Indian Journal of Marine Science, 8*, 94–97.

Volmer, D. A., Brombacher, S., & Whitehead, B. (2002). Studies on Azaspiracid biotoxins. I. Ultrafast high-resolution liquid chromatography/mass spectrometry separations using monolithic columns. *Rapid Communications in Mass Spectrometry, 16*, 2298–2305.

Wynne, T. T., Stumpf, R. P., Tomlinson, M. C., Warner, R. A., Tester, P. A., Dyble, J., & Fahnenstiel, G. L. (2008). Relating spectral shape to cyanobacterial blooms in the Laurentian Great Lakes. *International Journal of Remote Sensing, 29*(12), 3665–3672.

Yasumoto, T., Murata, M., Oshima, Y., Matsumoto, G. K., & Clardy, J. (1984). Diarrheic shellfish poisoning. In *Seafood toxins*. Washington, DC: American Chemical Society.

Chapter 2
Algal Blooms: Potential Drivers, Occurrences and Impact

Abstract The growth of marine phytoplankton (both non-toxic and toxic) is generally limited by the availability of nitrates and phosphates, which can be abundant in coastal upwelling zones as well as in agricultural runoff. The type of nitrates and phosphates available in the system is also a factor, since phytoplankton can grow at different rates depending on the relative abundance of these substances (e.g. ammonia, urea, nitrate ion). A variety of other nutrient sources can also play an important role in affecting algal bloom formation, including iron, silica or carbon. This chapter has given importance to gain insight into the characteristics of algal blooms along with their potential drivers in relation to the marine environment. The chapter has also highlighted the current understanding on the mechanisms of algal bloom and addresses the regional differences in the persistence and causative agents of algal bloom in eutrophic tropical aquatic systems.

Keywords Algal bloom · Dimethyl sulphide · Upwelling and downwelling · Allelopathy · Remote sensors · Water quality · Arabian Sea · Bay of Bengal

2.1 Introduction

Algal bloom is the natural phenomena due to sudden dense outburst of a group of microalgae, usually dominated by dinoflagellates or cyanobacteria (blue-green algae/bacteria), diatoms followed by raphidophytes and haptophytes, with low species diversity. This biotic stress results from a complex and cohesive mechanisms involving physico-chemical and biological processes. The water becomes noticeably coloured (green, brown, yellow, red or reddish brown) related to the high cell abundance of the key species concerned, mainly monospecific or single species, stretching from metres to hundreds of kilometres either at the surface or as thin subsurface layers of high cell concentration (as shown in Fig. 2.1). The blooms are rapid and have perceptible increase in microalgae biomass or standing crop, accounting 2–5 million cells/litre, even exceeding 10 million cells/litre in extreme cases. The frequency and severity of such blooms seem to be increasing during recent period especially in subtropical to temperate climates, getting access over the

S. K. Sarkar, *Marine Algal Bloom: Characteristics, Causes and Climate Change Impacts*, https://doi.org/10.1007/978-981-10-8261-0_2

Fig. 2.1 Red tide occurrence in Long Island (extends 190 km eastward), an expansive, densely populated island in southeastern New York State, USA. (Courtesy: Dr. Mindy L. Richlen, National Office for HABs, WHOI, USA)

coastal regions in recent times where no previous record of algal bloom was occurred.

The development and proliferation of algal blooms are favoured by a set of complex environmental factors including sudden influx of nutrients (through upwelling or tidal turbulence), physical factors (specific circulation and current patterns, wind, light intensity), ecosystem disturbance (stable/mixing conditions, turbidity) or washout of nutrients into sea from land sources. Frequency and magnitude of algal bloom are also linearly related to intensive rainfall that substantially help to increase the amount of particulate and dissolved organic matter (POM and DOM) in nearby coastal regions which has been recently asserted by Meng et al. (2016) in a hypereutrophic lagoon in Taiwan.

2.2 Important Drivers for Algal Bloom

2.2.1 Upwelling and Downwelling

Open oceans have vertical currents (recognized as upwelling and downwelling) in addition to the commonly known horizontal currents where winds cause surface waters to diverge (move away) from a region (causing upwelling takes place along much of the equator) or to converge towards some region (causing downwelling). Two fundamental types of upwelling are recognized: (1) wind-driven coastal Ekman upwelling and (2) equatorial divergent upwelling. Both these vertical currents are significantly important for abrupt increase of biological productivity where the greatest primary production takes place. The deep, nutrient-rich waters (nitrogen and phosphorus compounds, etc.) are delivered to the surface photic layer which triggers marine algae to multiply and produce the biomass that drives the marine food chain. An appropriate wind is needed to cause this upwelling while also ensuring temperature, and transport are also favourable for *C. polykrikoides* blooms (Kim et al. 2016).

Phenomenon of upwelling was studied by Anas et al. (2018) with the onset of southwest monsoon along the southwest coast of India. They had noted that the upwelling induces changes in the abundance and community structure of bacteria in the recurring mudbank in this coastal region. The hydrological variables were also simultaneously changed as follows: (1) a decrease in sea surface temperature and dissolved oxygen in bottom waters and increase in salinity and (2) nitrite and phosphate that showed a significant increase during upwelling conditions.

The world's most potential fisheries are located on coastal upwelling regions (especially in the eastern boundary regions of the subtropical gyres). In contrast, downwelling reduces biological productivity and transports heat, dissolved materials and surface waters rich in dissolved oxygen to greater depths. This occurs along the west coast of Alaska in the eastern boundary region of the Gulf of Alaska gyre (driven by winds in the Aleutian Low). Hubbart et al. (2012) observed that both upwelling and downwelling cycles drive high variability of toxigenic phytoplankton in the west coast of South Africa.

Strong algal blooms including harmful algal blooms (HABs) occurred in the Vietnamese coastal upwelling regions during southwest (SW) summer monsoon as evidenced from satellite images and in situ observations (Dippner et al. 2011). The blooms are triggered by excessive micronutrients provided by this period; nutrients are provided by upwelling along with river runoff from the Mekong River. The model produces realistic results compared to in situ observations and satellite images, hence important for real-time forecast beneficial for coastal protection.

High variability of toxigenic phytoplankton community and the high variability in cellular toxicity, and the consequent selective uptake and loss of toxins, and/or transformation of toxins by the mussel *Choromytilus meridionalis* due to upwelling-downwelling cycles drive off the west coast of South Africa (Hubbart et al. 2012). Rapid changes in phytoplankton assemblages revealed the following three groups of toxigenic phytoplankton: (a) the dinoflagellate *Alexandrium catenella*; (b) several species of *Dinophysis*, including *Dinophysis acuminata*, *Dinophysis fortii*, *Dinophysis hastata* and *Dinophysis rotundata*; and (c) members of the diatom genus *Pseudo-nitzschia*.

2.2.2 Intensive Rainfall

The effects of rainfall on coastal ecosystems have received great attention in the last few decades (Arhonditsis et al. 2002; Meng et al. 2015). Since extreme rainfall events are predicted to increase in the near future, the aftereffects are also likely to be more evident (Milly et al. 2002; Palmer and Ralsanen 2002; Power et al. 2013; Tan et al. 2015). There is a direct relationship between the rainfall intensity on algal blooms, which results drastic fall in salinity. The magnitude of phytoplankton blooms also increased linearly with increasing rainfall intensity due to substantial nutrient enrichment (dissolved inorganic nutrients) and dissolved organic matter. The event is caused by drainage or runoff water from adjacent regions that are channelled into the aquatic systems during rainfall events. However, there exists scant information on the effects of rainfall intensity over a longer periods of time in coastal ecosystems, especially in tropical regions (Staehr et al. 2012 and citations therein). Recently, Meng et al. (2017) opined that the frequency and magnitude of phytoplankton blooms may increase in the future due to possibility in increase of extreme rainfall. The scientists had evaluated the effect of rainfall intensity on phytoplankton blooms in a hyper-eutrophic lagoon in Taiwan.

2.2.3 Allelopathy

Allelopathy may be defined as the ability of certain harmful algal species or microorganisms to produce and release chemicals (allelochemicals) or compounds that inhibit their potential co-occurring competitors for similar resources. Production of these chemicals, mostly toxic in nature, interferes with growth and development of accompanying species, and, in the case that the growth of competitors is inhibited, it gives competitive advantages to the species which produces the allelopathic substances (Lewis 1986).

Moreover, these allelopathic/toxic substances seemed to be secondary metabolites with harmful capacity, and their production have been shown to be enhanced by stress factors, such as N and P unbalanced ratios at limiting concentrations (Granéli and Hansen 2006). The resulting stressed conditions initiate the production of allelochemicals to provide an advantage over potential competitors for the limiting nutrient (Fig. 2.2). This allelopathic potency was demonstrated for bacteria (Long and Azam 2001), cyanobacteria (Suikkanen et al. 2005), dinoflagellates (Tillmann and John 2002; Kubanek et al. 2005), as well as in diatoms in coastal environments (Subba Rao et al. 1995).

Eutrophication alters the nitrogen-to-phosphorus balance. Algal species that can compete successfully for available growth-limiting nutrient(s) have the potential to become dominant and form blooms. In some algae, the stress conditions imposed by the shifted nutrient supply ratios can stimulate production of allelochemicals that inhibit potential competitors. Thus, under cultural eutrophication, altered nutrient (N, P) ratios and limiting nutrient supplies can stimulate increased production of allelochemicals, including toxins, by some algal species and accentuate the adverse effects of these substances on other algae.

Phytoplankton allelopathy was possibly first observed by Harder in 1917 (Wolfe and Rice 1979) when he reported auto inhibition in the freshwater cyanobacterium *Nostocpunctiforme*. The International Allelopathy Society defined allelopathy as any process involving secondary metabolites produced by plants, algae, bacteria and fungi that influence the growth and development of agricultural and biological systems.

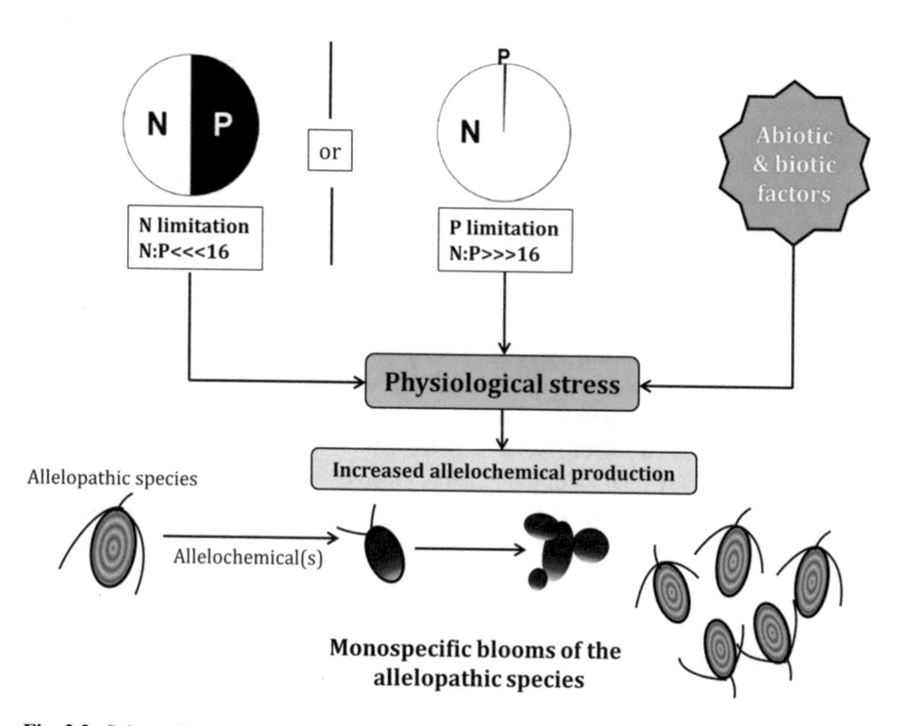

Fig. 2.2 Schematic representation of algal bloom occurrence due to allelopathic phenomenon

Harmful algal blooms (HABs) have exponentially spread across the globe covering fresh, estuarine and coastal marine waters (Allen et al. 2006) (details of HABs have been given in Chap. 3). Some of these blooms revealed worse changes in comparison to others, for example, the devastating bloom of the planktonic green alga *Chrysochromulina polylepis* in Scandinavia occurred in 1988 that killed all fauna and flora across ~ 75,000 km^2 area (Granéli and Hansen 2006). Excessive inputs/influx of N and P generally results high-biomass blooms, and their respiration and final decomposition lead to oxygen depletion in the water column. All such changes in water quality have severe deleterious effects on benthic community, resulting low benthic productivity (Glibert et al. 2005). Some algal blooms, including some high-biomass blooms, also involve species that produce toxins and other secondary metabolites which are released into the water. It has been hypothesized that the primary purpose of these chemicals may be to inhibit the growth of other competing phytoplankton species (allelopathy) and to decrease losses by killing their grazers (grazer deterrence).In freshwaters, most allelopathic species are cyanobacteria, whereas in marine and estuarine waters most are flagellates, especially dinoflagellates and haptophytes. The most extensive effects on aquatic ecosystems have been associated with allelochemicals produced by flagellates (Legrand et al. 2003).

The strongest allelochemicals possess haemolytic capacity aquatic systems, which disrupts the cell membranes of other competitors and eventually kill them in the process. However, algal allelochemicals are not so fatal, might only hinder the ecophysiological functions in the target species for a short time period, not causing death. For example, photosynthesis inhibition, decrease in growth rate or grazing inhibition are some important features observed to a certain level (Legrand et al. 2003). In addition, induced cyst formation may be another non-lethal reaction to allelochemicals (Legrand et al. 2003). Allelopathic compounds released by some phytoplankton species seem to be effective only for a relative short period of time. For example, cell-free filtrates of *Prymnesium parvum* added to cultures of *Thalassiosira weissflogii*, *Rhodomonas* cf. *baltica* and *Prorocentrum minimum* had significant adverse impact on cell numbers, but the exposed species began to recover within a few days (Fistarol et al. 2005). Under stressed conditions, allelopathic algae have the capability to use the limiting nutrient resources by eliminating or inhibiting the growth of their competitors via increased production and liberation of allelochemicals. Cultural eutrophication due to nutrient enrichments is an underlying factor developing many blooms of allelopathic algae by altering the N:P balance. The stressed conditions promote the production of allelochemicals to afford an advantage over potential competitors for the limiting nutrient (Granéli et al. 2008).

2.3 Algal Bloom and DMS Production

Algal bloom dynamics, especially their speciation and growth phases, seemed to play significant roles in shaping the distribution of dimethyl sulphide (DMS) firstly through variable biosynthesis of Dimethylsulphoniopropionate (DMSP) by diverse members of phytoplankton communities (Keller 1989; Matrai and Keller 1994). Both DMS and DMSP are methylated sulphur compounds, and interest in these compounds are increasing and diversifying due to their ecological, biochemical and biogeochemical implications. Dimethylsulfide (DMS) is the largest natural source of sulphur to the atmosphere and may influence climate regulation by affecting atmospheric chemistry and the heat balance of the atmosphere (Gypens and Borges 2014; Charlson et al. 1987) though the significance of this feedback still remains uncertain (Carslaw et al. 2010; Quinn and Bates 2011). DMS is the most abundant volatile sulphur compound in the surface ocean and represents the major natural source of reduced sulphur to the global troposphere (Andreae and Crutzen 1997). This is mainly produced by the enzymatic cleavage of its biological precursor dimethyl sulfonio propionate (DMSP), an abundant and widespread intracellular compound found in marine microalgae (Keller 1989) and in other halophytic plants. DMSP is mainly produced by a limited number of marine microalgae, haptophytes and dinoflagellates being usually characterized by higher cellular DMSP concentrations than diatoms (Keller 1989; Stefels et al. 2007).

DMSP production by phytoplankton displays a wide range of variations among the phytoplankton taxa, from non-detectable (e.g. few cyanobacteria and diatoms) to a large amounts within dinoflagellates and prymnesiophytes group (Keller 1989). This trend of large variability, both across taxonomic groups (Keller 1989) and within taxa, is related to other environmental conditions (Stefels et al. 2007). DMSP production is also modulated by other physico-chemical factors, such as intensity of solar radiation, nutrient repletion or depletion, oxidative stresses and changes in salinity or temperature.

Dimethylsulphoniopropionate (DMSP) is an organic sulphur compound synthesized mainly by marine macro- and microalgae and of significant ecological interest as it acts as the principal precursor of the climatically active gas dimethyl sulphide (DMS) in the ocean cycle. Archer et al. (2001) presented an elaborate account of the rates of production of particulate DMSP (DMSPp) and turnover by microzooplankton in surface waters of the northern North Sea. They observed that the phytoplankton communities were characterized by DMSP-rich taxa such as *Emiliania huxleyi* (Lohman) and *Prorocentrum minimum* (Pavillard) J. Schiller and DMSPp/chlorophyll *a* (chl *a*) ratios of 64–162 nm μg^{-1}. *E. huxleyi* is a coccolithophore cosmopolitan species, whereas *P. minimum* is a red tide dinoflagellate.

Perusal of literature reveals that coastal environments are hotspots of DMS emissions and will strongly respond to eutrophication in addition to ocean acidification at decadal time scales. The major anthropogenic disturbance of marine ecosystems can lead to a stronger response of DMS emissions that actually counters the effect of ocean acidification in a case study in the strongly eutrophied Southern Bight of the North Sea (SBNS). This is a hotspot of DMS emissions (Uher 2006; Lana et al. 2011) because of the occurrence of intense blooms of *Phaeocystis* that are high DMSP producers. The response of DMS emissions to eutrophication at global scale cannot be modelled at present because the resolution of general ocean circulation models or earth system models is insufficient to represent coastal areas influenced by rivers.

The modification of microalgae dominance and production in response to climate warming (e.g. Blanchard et al. 2012), eutrophication (e.g. Mackenzie et al. 2011) or ocean acidification (e.g. Doney et al. 2009) is expected to change DMS emissions in the future with a potential positive or a negative feedback on climate change (Bopp et al. 2003; Gabric et al. 2004; Kloster et al. 2007; Vallina et al. 2007; Cameron-Smith et al. 2011; Six et al. 2013).

2.4 Algal Bloom in Tropical Coastal Regions: Case Studies

2.4.1 Kalpakkam Coastal Regions, Southeast Coast of India

The section highlights the subsequent negative impacts due to occurrence of algal blooms in different tropical coastal regions (as depicted in Fig. 2.3) along with the causative bloom-forming species. Each region is characterized by different microalgal groups that have their unique adaptive strategies to offset environmental disadvantages. As evidenced from the following case studies, the occurrence of algal bloom is related to complex environmental conditions being influenced by several physical factors such as tidal currents, salinity, light intensity, stratification, and

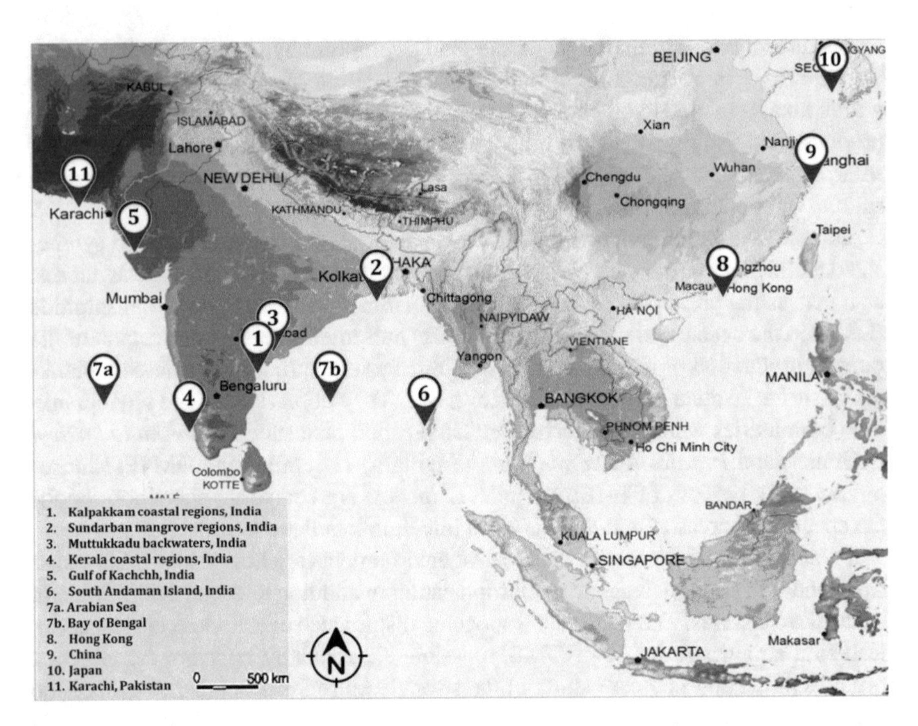

Fig. 2.3 Map showing the occurrence of algal blooms in different coastal regions in India and other tropical coastal regions

sudden influx of nutrients as well as ecosystem disturbance. Kalpakkam coast (12°33′ latitude and 80°11′E longitude) is situated about 68 km south of Chennai megacity, Tamil Nadu, on the east coast of India. This is bounded with Buckingham Canal on the western site and the Bay of Bengal on the eastern site. The coast in this region is relatively flat, and the foreshore is primarily composed of coarse canal. The coastal currents at Kalpakkam has seasonal character, and during SW monsoon the current is northerly (February–October) with a magnitude of 0.2–1.8 km h^{-1}, and during NE monsoon the current is southerly (October—February) with a comparative low magnitude of 0.1–1.3 km h^{-1}.

Based on the rainfall pattern and associated hydrological changes along the Kalpakkam coast, the whole year can be divided into three seasons, namely, (a) summer (February–June), (b) southwest monsoon (July–September) and (c) northeast monsoon (October–January). This coastal region of peninsular India receives bulk of rainfall (~72%) from northeast monsoon and the coverage annual rainfall in ~1200 mm. Kalpakkam is significant since it hosts the nuclear power plant, Madras Atomic Power Station (MAPS), consisting of two pressurized heavy water reactors (PHWRS). The coastal ecosystem is partly influenced by inputs from river polar backwater, viz. Sardar and Edaiyur, that transport effluents from urban settlement and industries. Moreover, Buckingham Canal runs parallel to the coast carrying urban sewage, pesticides and fertilizer received from agricultural activities and the salt pan industries intersect the backwater. During the monsoon season, especially in NE monsoon, these two backwaters are opened to the coast, discharging considerable amount of freshwater to the coast for 2–3 monsoonal period.

Bloom of *Trichodesmium erythraeum* (Ehr.) and its impact on water quality and plankton community structure in the southeast coast of India was studied by Mohanty et al. (2010). Contribution of *Trichodesmium* to the total phytoplankton density ranged from 7.79% to 97.01% with the highest density of 2.88×10^7cells l^{-1}. The population density of phytoplankters ranged from $1.23 \times 10^5$and 2.94×10^7cells l^{-1} showing more than two-order increase during the peak bloom period. The cell density was found to surpass all the earlier reported densities from east and west coast of India. The cyanobacterial bloom had a considerable impact on the zooplankton community of the coastal waters. A significant reduction in zooplankton density was observed from bloom to post-bloom periods.

Trichodesmium erythraeum, a prolific nitrogen-fixing diazotrophic marine cyanobacterium, is a potent bloom-forming species recorded in tropical and subtropical waters, particularly in the eastern tropical Pacific and Arabian Sea. The large blooms are widely recognized as 'sea sawdust' contributing >30% of algal blooms of the world. Estimated global nitrogen fixation by *Trichodesmium* bloom (~ 42 Tg N year^{-1}) and during non-bloom conditions (~ 20 Tg N year^{-1}) suggests that it is likely to be the dominant organism in the global ocean nitrogen budget (Westberry and Siegel 2006; Capone et al. 1997). The species normally occurs in macroscopic bundles or colonies, and blooms are often extremely patchy in nature, mainly concerned to the physical variability of the water body (Kononen and Leppänen 1997). Perusal of literature revealed frequent occurrence of *Trichodesmium* blooms in Indian coastal regions, more frequently in the west coast (Qasim 1970; Sarangi et al. 2004; Prabhu et al. 1965; Devassy et al. 1978; Devassy 1987; Shetty et al. 1988; Koya and Kaladharan 1997; Krishnan et al. 2007) as compared to east coast (Jyothibabu et al. 2003; Ramamurthy et al. 1972; Santhanam et al. 1994a, b; Satpathy et al. 2007). Equipped with buoyancy regulating gas vesicles and nitrogen fixation enzymes, *Trichodesmium* is considered as an organism well adapted to stratified, oligotrophic conditions (Capone et al. 1997).

Occurrence of *Trichodesmium* bloom from east and west coast of India have been recorded far away from the coast (> 30 km). This relates to the second report of bloom formed by this species which occurred almost adjacent to the coast similar to previous year report from the same locality (Satpathy et al. 2007). During a regular coastal water monitoring programme, a prominent discolouration of the surface water was noticed in the coastal waters of Kalpakkam coastal waters on February 19, 2008. The bloom was very dense and created yellowish-green-coloured streaks of about 4–5-m-width and 10–20-m-long patches. The entire bloom extended to several kilometres along the coast. The phytoplankton responsible for discolouration was identified as *Trichodesmium erythraeum.* Nevertheless, blooms of *Noctiluca scintillans* (Sargunam et al. 1989), *Asterionella glacialis* (Satpathy and Nair 1996) and *Trichodesmium erythraeum* (Sargunam et al. 1989) in the coastal waters of the Kalpakkam have been reported; the present one has many interesting features. The bloom patches of *A.glacialis* recorded in Kalpakkam coastal regions

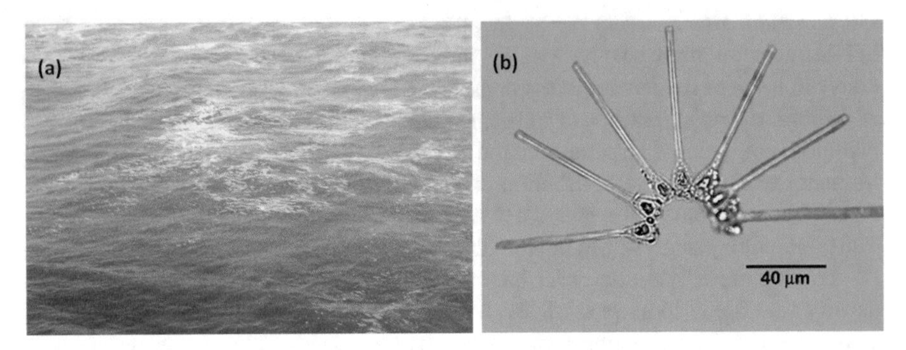

Fig. 2.4 (**a**) *Asterionella glacialis* bloom patches observed in the coastal waters of Kalpakkam, India, during January 17–21, 2015, and (**b**) a chain of *A. glacialis* observed under light microscope. (Courtesy: K.K. Satpathy, Environment Division, IGCAR)

along with a chain of the same species have been given in Fig. 2.4. Although the data collected during our regular work were not concerned directly with an investigation into the causes of the bloom, the interest stimulated from the studies of various physico-chemical and biological characteristics of the coastal water justifies the purpose of this paper. The acumen in investigating *Trichodesmium* bloom appearance and distribution stems from the report regarding its harmful nature, sometimes causing damages to coastal fish and shellfish fauna (Bhat and Verlencar 2006). Thus, studying the causes that favour the appearance of such bloom has social and economical connotations. The impact of bloom on coastal water quality and phytoplankton community is reported in this paper along with the characteristic feature of the bloom.

Trichodesmium is considered as an organism well adapted to stratified, oligotrophic environment. Hence, its abundance should be high in the boundary currents and decrease towards the coast in the nitrogenous enriched environments. So far, no report has been documented for *T. erythraeum* bloom in the coastal zone near the coastal areas (within 600 m from the shore). Results revealed that the bloom constituted both individual trichomes and colonial forms took part in forming the bloom (Fig. 2.5), although the later dominated about of 80–90%. The trichome length varied a wide range from 300 to 1200 µm. Unlike the west coast of India wherein algal bloom is generally restricted during the beginning of SW monsoon period (May–September), the present bloom in Kalpakkam was noted during the end of NE monsoon period. The study was undertaken during the transition period when the coastal water current seemed to change from southerly to northerly direction. The lowest magnitude of current is pronounced during this lull phase at this location. Blooms have been reported to be conspicuous in calm conditions. The prevalent calm condition facilitates the trichomes to form dense rafts on the surface. Phytoplankton community showed a distinct variation from qualitative as well as quantitative point of view during the study. Out of 69 phytoplankton species, diatoms were dominant (62), followed by 5 dinoflagellates, 1 silicoflagellate and the cyanobacterium *Trichodesmium erythraeum*. The numerical density of phytoplank-

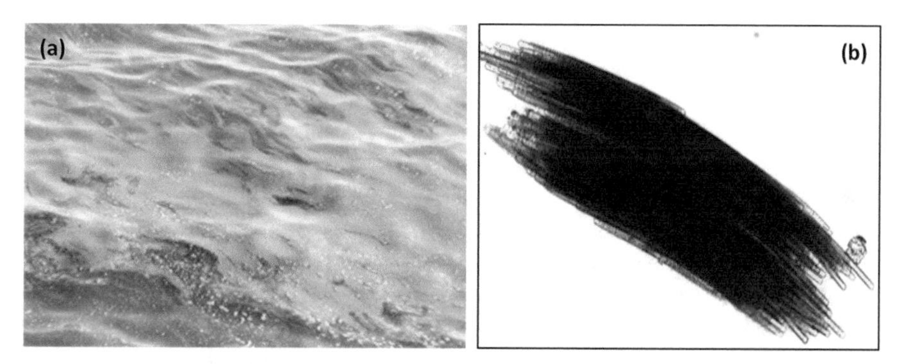

Fig. 2.5 Discolouration of coastal waters of Kalpakkam by (**a**) *Trichodesmium erythraeum* bloom patches on February 19, 2008; (**b**) magnified view of bundles formed by trichome. (Courtesy: K.K. Satpathy, Environment Division, IGCAR)

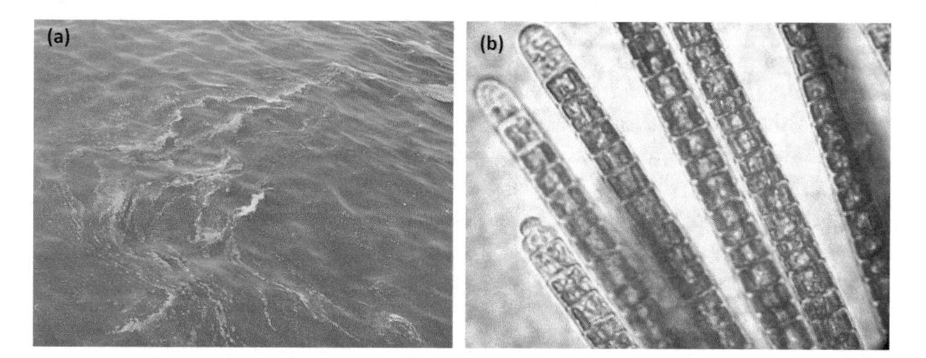

Fig. 2.6 A photographic view of (**a**) phytoplankton bloom on March 16, 2007 by *Trichodesmium erythraeum* in the coastal waters of Kalpakkam, east coast of India, and (**b**) the photomicrograph of the *Trichodesmium* filament (×2400). (Courtesy: K.K. Satpathy, Environment Division, IGCAR)

tons showed variations from 1.23×10^5 to 2.94×10^6 cells l^{-1} showing more than twofold increase during the peak bloom period; in contrast, the lowest cell density was recorded during post-bloom period. Most strikingly, *Trichodesmium* was present exclusively during the bloom period and was totally absent during the pre- and post-bloom observations. Contribution of *Trichodesmium* to the total cell count ranged from 7.79% (1.10×10^4 cells l^{-1}) to 97.01% (2.88×10^7cells l^{-1}). It is reported that the species is more abundant in subsurface layers (20–30 m) as compared to surface water (Subba Rao 1969). *Trichodesmium* density as recorded in Kalpakkam coastal regions was found to be significantly higher than the previously reported value of 3.38×10^6 cells l^{-1} recorded by Ramamurthy et al. (1972) and 4.80×10^6 cells l^{-1} by Krishnan et al. (2007). The photographic view of algal bloom caused by *T. erythraeum* and the filament of the species has been given in Fig. 2.6. Perusal of published literature revealed that the study in the density of *Trichodesmium* is the maximum from Indian waters recorded so far and is about 1.75 times higher from the highest value (1.75×10^7 cells l^{-1}) reported by Santhanam et al. (1994a, b) from Tuticorin Bay, Tamil Nadu. Community structure of phytoplankton showed that the number of species on a single observation varied between 7 species (on the day of highest cell count) and 24 during the post-bloom period. Relatively less number of species was recorded during the bloom as comparison to pre- and post-bloom periods which corroborated the previous report by Mishra et al. (2006) during *Asterionella* bloom. Interestingly, number of species was greater in the morning collection than afternoon. It is worth to mention that *Trichodesmium erythraeum* can tolerate relatively high amount of irradiance in the surface water during afternoon as compared to morning period. However, species not tolerant to irradiance evade the surface water leading to low species diversity. Based on the numerical abundance, 30 species were showed 74.19–99.70% of the population density. Out of these, *Asterionella glacialis*, *Nitzschia longissima*, *Thalassionema nitzschioides*, *Thalassiosira decipiens* and *Thalassiothrix longissima* were overall present during the study period. Species such as *Biddulphia heteroceros*, *Cocconeis distans* and *Leptocylindrus minimum* were found only during the pre-bloom period and totally absent during bloom and post-bloom periods. On the contrary, two species of *Biddulphia* (*B. aurita* and *B. rhombus*) were found only during the post-bloom period. This clearly indicated that presence of *Trichodesmium erythraeum* favours growth of a selected group of diatoms during the post-bloom period. Distinct variations in all the three diversity indices were noticed during the study period.

All the ecological indices are of relatively high values during the pre- and post-bloom periods justifying that the phytoplankton community was floristically rich during these periods. A significant decrease in diversity indices was concomitant on the day of the maximum *Trichodesmium* density which is due to exclusive dominance of this species and the presence of traceable number of residual phytoplankton species.

Phytopigments exhibited large variations with Chl a and phaeophytin ranging from 1.21–42.15 mg m^{-3} to 0.78–46.23 mg m^{-3}, respectively. The maximum concentration was recorded during the bloom which coincided with highest cell density. Concentration of chlorophyll *a* and phaeopigments showed high during bloom as compared to pre- and post-bloom periods which were about 20 times higher than the normal values. However, concentrations of these pigments were relatively high during the post-bloom period as compared to the pre-bloom period. These typical successional phases endorses that phytoplankton growth gradually increased from post-monsoon to summer in this coastal region of Bay of Bengal (Koya and Kaladharan 1997; Krishnan et al. 2007). The present observation related to unusually high pigment concentrations was corroborated by Ramamurthy et al. (1972), Pant and Devassy (1976) and Satpathy et al. (2007) during *Trichodesmium* bloom and Mishra et al. (2006) and Mishra and Panigraphy 1995during *Asterionella* bloom. During *T. erythraeum* bloom in Kalpakkam coastal regions at a depth of ~8 m, relatively high temperature, low current magnitude, stable salinity (~ 33 psu) and low nitrate concentration along with high ammonia level were observed in the ambient medium. The cell density was found to surpass all the earlier reported densities from east and west coast of India. Continuous and systematic monitoring of physico-chemical parameters on a long-term basis is recommended which would help in comprehending the cause and remedial measures for mitigating these toxic bloom phases.

2.4.2 *Sundarban Mangrove Wetland, India*

Indian Sundarban (21°13' to 22°40'N and 88°03′ to 89°07′E), the Anthropocene tide-dominated megadelta covering area of a 9630 km^2, is equipped with continuous single tract of mangrove forest comprising of about 28 major species. This is considered as a 'global biodiversity hotspot', inhabited by numerous species of plankton (phytoplankton and zooplankton), benthic organism (meio-, micro- and mega-benthos), fishes, amphibians and mammals (Gopal and Chauhan 2006).

Situated in a tropical, macrotidal, low-lying coastal zone, it is formed at the confluence of Ganges (Hooghly) River and the Bay of Bengal (BOB). The region is under severe pressure from anthropogenic activities and discharges from domestic, industrial, agricultural and aquacultural runoff leading to enrichment of nutrients and inorganic and organic pollutants. It is worth to refer that only 20% of the waste generated in 42 riverside municipalities is treated at sewage treatment plant, and the rest is discharge directly in the Hooghly River resulting negative consequences on biodiversity, productivity and water quality as a whole. Due to frequent occurrence of natural catastrophic events (such as cyclones, tsunami, flooding), this could be considered as the vulnerable coastal zone.

A dense monospecies bloom caused by the solitary species centric diatom *Hemidiscus hardmannianus* (Greville) Kuntze 1898 (Bacillariophyceae) was recorded in western part of Indian Sundarban coastal regions on July 22, 2010. The species was present in abundance, and the mean density reached as high as 8.86×10^6 cells/L, when water coloration turned green with an obnoxious smell from water surface. As a consequence there were significant adverse effects on plankton. The diversity as well as density of phytoplankton, microzooplankton and mesozooplankton community showed poor values during the bloom period which was persisted for 1 week. This is an open marine, large diatom (mean diameter 0.8 mm) and most common in warm water. Occurrence of such algal blooms with their subsequent negative impact on biotic community certainly deserves special attention to adopt adaptive ecofriendly management strategies.

2.4.2.1 Water Quality Characteristics Due to Single Species Blooming

Water quality characteristics exhibited pronounced changes due to the occurrence of bloom when the temperature was comparatively high (ranged 27.5–33.28 °C), and such ambient temperature has already been considered as significant factor that initiates the bloom phase of *H. hardmannianus*. Salinity showed wide range of variation (7.1–29.8), and a steady increase in salinity was recorded from the upper limnetic zone of the estuary (Lot 8) towards the open ocean (Gangasagar; S_3). The maximum and minimum value of dissolved oxygen (DO) were reach as 5.68 mg/L during bloom and 3.51 mg/L during post-bloom period. Relatively high dissolved oxygen level was recorded during bloom condition which may be attributed release of O_2 by the dense diatom biomass, and this phenomenon was also observed by Satpathy et al. (2007). During post-bloom period, low dissolved oxygen concentrations was recorded pointing out algal biomass is in decaying stage and the bacteria concerned for the decay process are mainly consuming most of the dissolved oxygen. Keeping pace with low DO concentration (~3.67 to 0.24 mg/L), the BOD value was comparatively high (~1.92 to 0.67 mg/L) when the ambient medium was almost hypoxic/anoxic condition. During the peak bloom phase, siltation rate in the surface water was significantly low leading to unusual lower values of turbidity (1.6–0.29 NTU). In contrast, turbidity showed extremely high value during post-bloom condition (13.66–2.57 NTU), when the diatom stock sedimented at the bottom as dead mass. Subramanian and Purushothaman (1985) also recorded similar lower turbidity values of surface water during *H. hardmannianus* bloom. Chlorophyll a (chl a) also showed wide range of variations (0.8–8.9 mg/m^3) exhibiting an exponential increase (8.64 mg/m^3) coinciding with the highest cell density (8.86 × 106 cells/L). In general, a comparatively high chlorophyll a concentration was reasonably high during bloom as compared to pre- and post-bloom period, and the peak value of the pigment was eight to nine times higher than the non-bloom period. Satpathy et al. (2007) and Mishra et al. (2006) also recorded similar unusually high pigment concentration during bloom phases from eastern coastal part of India. The prevalent enrichment of chlorophyll level (4.0–8.0 mg/m^3) was also distinct from the chlorophyll images as derived from satellite-derived data. TMI satellite-derived sea surface temperature (SST) and wind speed maps have been interpreted. The SST is observed to be warmer (28–32.8 °C) and wind speed in the range of 5–7 m/s. Hence it is assumed that the diatom *H. hardmannianus* bloom was favored by higher temperature and moderate wind speed.

2.4.2.2 Changes in Plankton Community

During bloom phase, phytoplankton community in terms of species richness and evenness exhibited pronounced variation in its qualitative as well as quantitative aspects. Out of 61 identified species, diatom, dinoflagellates and cyanobacteria were consisted of 55, 4 and 2, respectively. An overwhelming dominance of the bloom-causing diatom *H. hardmannianus* was encountered contributing ~100% of the total phytoplankton. The bloom-causing species *H. hardmannianus* appeared as a sudden spurt in cell number when the phytoplankton population density increased abruptly from 12.84 × 103 to 8.86 × 106 cells/L resulting in >1000 times increase during bloom. The prevalent density of the species was remarkably higher than the previous value of 49 × 10^3 cells/L reported by Subramanian and Purushothaman (1985) in Parangipettai coast, southern part of India and by Shetty and Saha (1971) in coastal waters of Sundarban wetland. A bloom may occur as an isolated community event or co-occur with other species, some of which may then reach population maxima several orders of magnitude higher (Smayda 1997). Saunders and Glenn (1969) considered a species to be a bloom former when its density at any one time is enough to give a total surface area of 107 mm^2/L. Accordingly, Santhanam (1976) calculated the surface area of a single cell of *H. hardmannianus* as 117,068 mm^2 and pointed out that 85 cells/L were required to comprise a bloom. The surface area has been measured as 187,043 mm^2, and as per above reference, only 54 cells are required to form a bloom which persisted for a week period and could be categorized as a periodic event (Takahashi et al. 1977). The extremely high numerical density along with large surface area of this diatom as observed around certainly formed a bloom (Fig. 2.7).

During the bloom condition, an improvement of tintinnids (planktonic ciliate) population density was related to extreme high-density pressure of *H. hardmannianus*. This characteristics ciliate community regained the biomass when this bloom-causing diatom disappeared allowing the normal phytoplankton community

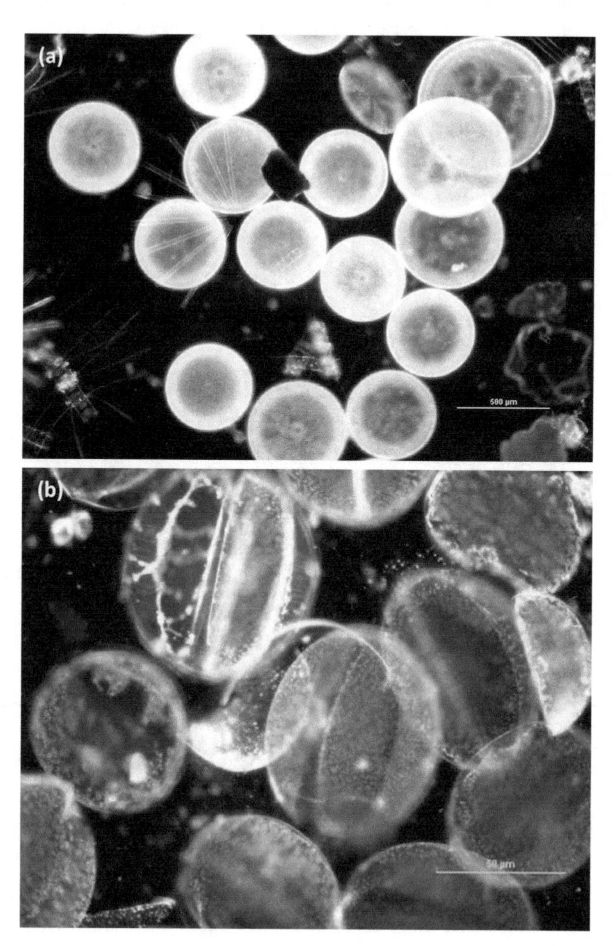

Fig. 2.7 Microphotographs of (**a**) multispecies diatom bloom and (**b**) *Hemidiscus hardmannianus* bloom in Sundarban mangrove wetland

structure to be established. However, a contrasting feature was also reported from Great South Bay, New York, by Duguay et al. (1989) which recorded a reverse scenario during a brown tide outbreak, and it is interpreted that the food condition was favourable for these tiny organisms. Mesozooplankton composition was very unusual. During the blooming phase, mesozooplankton composition was very unusual as compared with the pre- and post-blooming phase. Like Tintinnida, as discussed above, the greatest impoverishment of mesozooplankton was also noticed during blooming period when very low density of nauplii, zoea and *Lucifer* sp. (52.91, 62.91 and 15.01 ind./m^3, respectively) was recorded. In contrast, during pre-bloom period, the average numerical abundance of MZ was 1086 ind./m^3 in which copepod showed dominant part (~76.94%) of the total zooplankton. Copepods shared the most dominant component during pre- and postbloom periods at all three study sites, especially the suborder Calanoid group contributing about 60.0–97% of total mesozooplankton. Predominance of calanoid copepods in marine and estuarine waters under normal conditions is well established worldwide including Sundarban coastal waters (Sarkar et al. 2007). However, a total contrast feature was noticed during the bloom period when copepods were either absent or present with few species. The chaetognath, represented by the solitary species *Sagitta bedoti*, was also not present during the bloom condition. For meroplankters, polychaete larvae appeared in considerable numbers during the bloom period. Decapod larvae (comprising of *Penaeus monodon*, *P. indicus*, *Metapenaeus monoceros*, *M. dobsoni* and *M. affinis*) were found in considerable numbers only during pre- and post-bloom conditions.

2.4.2.3 Impact on *Hilsa* Fish Catch

A coincidence of large shoals of *Hilsa*, commercially important fish (Clupeidae family) to the bloom of *H. hardmannianus*, was recorded, and this was previously reported by Shetty and Saha (1971) from Sundarban. The huge number of diatom could possibly act as an indicator species to the large-scale inland shoaling of the fish. The large shoals of Hilsa might be related to the availability of optimal conditions at the edge of the diatom patches which act as barriers to *Hilsa* migration, as demonstrated by Savage and Wimpenny (1936) and a potential food item of this fish in the Herring family.

2.4.2.4 Water Quality Changes Due to Multispecies Blooming

Multispecies bloom caused by centric diatom (*Coscinodiscus radiatus*, *Chaetoceros lorenzianus*) and the pennate diatom (*Thalassiothrix frauenfeldii*) in the coastal regions of western part of Indian Sundarban was recorded by Biswas et al. (2014). This mangrove ecosystem of the Indian subcontinent is well known not only for the aerial extent but also for its species diversity; Sundarban mangrove environment is provided with numerous species of phytoplankton, zooplankton, microorganisms, benthic invertebrates, molluscs, amphibians and mammals (Gopal and Chauhan 2006). *Coscinodiscus radiatus* was the most dominant species throughout the year, sharing a good percentage of the phytoplankton community. Along with these two diatoms, there was a profuse bloom resulting in discolouration of water, viz. *Thalassiothrix frauenfeldii*, which is common in both temperate and tropical regions, and *Chaetoceros lorenzianus*, the chain-forming marine diatom. Water was deep greenish in colour due to multispecies diatom bloom. Documentation of this sporadic high abundance, together with significant species richness of the diatoms, requires more intensive systematic studies to protect the coastal environment of Sundarban.

Hydrographical parameters exhibited pronounced variations in three phases of diatom bloom, viz. pre-bloom, bloom and post-bloom condition. Surface water temperature showed low value of ~22 °C, and the minimum and maximum values of dissolved oxygen (DO) were recorded during post-bloom and bloom phases, respectively. A relatively high DO concentration was recorded during bloom period which might be attributed to photosynthetic release of oxygen during photosynthetic process of one dense algal mass which was also observed by other researchers (Satpathy et al. 2007). Turbidity value was extremely reduced (2 NTU) during peak bloom phase due to presence of large diatom mass in the surface layer. This increased moderately (9 NTU) during post-bloom phase, as the diatom stock usually sedimented at the bottom as dead mass. Nitrate and silicate exhibited maximum (nitrate = 27.52 μmol l^{-1}; silicate =71.98 μmol l^{-1}) values during post-bloom and minimum (nitrate = 13.62 μmol l^{-1}; silicate = 34.13 μmol l^{-1}) values during bloom phases. However, the phosphate concentration reached its maximum (0.94 μmol l^{-1}) during pre-bloom phase. Along with physical and chemical processes, phytoplankton uptake and replenishment by microbial decomposition of organic matter is a significant factor affecting phosphate concentrations in coastal waters (Satpathy et al. 2007).

Besides physical and chemical processes, phosphate concentrations in coastal waters mainly depend upon phytoplankton uptake and replenishment by microbial decomposition of organic matter (Satpathy et al. 2007). Hence, an overall reduction in nutrients is pronounced due to occurrence of bloom, which supported a rise in phytoplankton productivity in the area in relation to the pre- and post-bloom phases (Sasmal et al. 2005). Chlorophyll a (chl a) showed wide range of variations (0.40–4.70 mg m^{-3}) with an exponential increase (4.70 mg m^{-3}) coinciding with the highest cell density (11.4×10^5 cells l^{-1}). Concentration of chlorophyll remained generally high during bloom compared to pre- and post-bloom phase, and the peak value of the pigment was 4–5 times higher than the non-bloom phase. Remarkable high phytopigment concentrations have been previously reported by Satpathy et al. (2007) and Mishra et al. (2006) during diatom bloom phases in the eastern coastal part of India. Satellite remote sensing-derived chlorophyll images have been retrieved from the MODIS-Aqua sensor data which also endorsed enriched concentration of chlorophyll (5.0–6.0 mg m^{-3}) in this coastal region.

2.4.2.5 Plankton Community Structure

The phytoplankton community were highly affected qualitative and quantitatively due to the occurrence of the bloom. Out of 30 diatom species encountered from this study site, 18 species were observed during the bloom phase with superb dominance of *C. radiatus* (6.15×10^5 cells l^{-1}) along with *T. frauenfeldii* (2.07×10^5 cells l^{-1}) and *C. lorenzianus* (1.68×10^5 cells l^{-1}). These three species together comprised approx. 86% of the total community to form a severe diatom bloom, persisted for a week and could be categorized as a periodic event. They were common at the mouth of the Hooghly river estuary, and the total phytoplankton population density showed multifold increase from pre-bloom to bloom phase. The population density decreased to 9.6×10^4 cells l^{-1} in the post-bloom period when only 16 species were present. The result of ANOVA also revealed significant variations ($F = 3.86$, $p < 0.05$) between different bloom phases in relevance with phytoplankton abundance.

The microzooplankton community was represented exclusively by *Favella ehrenbergii* during the bloom phase. Barría de Cao et al. (2005) reported similar impoverishment of tintinnid species from Bahia Blanca estuary, Argentina, during algal bloom. The prevalent poor abundance is mainly due to scarcity of adequate food related to the cell size of phytoplankton prey, and this corroborated to the previous record from the Sundarban coastal regions (Biswas et al. 2013). A small number of tintinnids provided with large peristome and lorica oral diameter can graze on phytoplankton during the bloom period. Again, this bloom was constructed by chain-forming and projection-bearing diatoms of the genus *Thalassiothrix* and *Chaetoceros*. Thread extrusion enlarged phytoplankton cell size, making it difficult to be ingested by some tintinnids (Verity and Villareal 1986). A drastic fall in microorganism population during *Coscinodiscus* species bloom was also reported from Brazilian waters by Fernandes et al. (2001) and Naz et al. (2012) as well as from Port Blair, Andaman Island, by Elangovan et al. (2012). This change could be due to the inability of microzooplankton to graze on large-cell phytoplankton of harbour sample. A trophic coupling may exist between the other seasonal associations and a more diverse phytoplankton, which includes short peaks of phytoflagellates, principally during late spring and summer (Gayoso 1999). However, the association of a *Chaetoceros* bloom with high concentrations of three tintinnid species (*Helicostomella subulata, Tintinnopsis parvula* and *Mesodinium rubrum*) was reported from Departure Bay, British Columbia (Margolis 1993). The species richness increased from 1.09 to 1.94 during pre-bloom and bloom period, respectively, since the species *F. Ehrenbergii* shared the major part to the total algal abundance during bloom phase. A reduction in species diversity (1.09–0.004) as well as species evenness (0.79–0.003) was noticed as the number of species reduced drastically during the bloom phase.

2.4.3 *Muttukadu Backwaters, India*

Prasath et al. (2014) observed a prolific cyanobacterial bloom caused by the toxic *Microcystis aeruginosa* (Kützing 1846) in eutrophic Muttukadu backwaters, southeast coast of India. The backwater extends inwards up to 20 km from the mouth and is normally cut off from the sea due to the formation of sandbar during May–September. During the winter season, the sandbar gets eroded, due to the inundation by the freshwater from the upper reaches, and the connection with the sea is restored. The depth of the estuary varies along the channel from 2 m in the middle of the channel to 1 m or less in most of the region and is 800–1050 m in width.

The bloom of blue-green algae, an ecologically and economically important cosmopolitan species among the planktonic cyanobacteria, affects the aquatic organisms and is of global concern. The blooms of *Microcystis* typically flourish in warm, turbid and slow-moving waters, and the highest biomass occur in eutrophic waters (waters enriched with nitrogen and phosphorous). The toxic effects exerted by the cyanobacterium bloom *M. aeruginosa* (average density of 6×10^8/L) have resulted in mass mortality of fishes. This alga is capable of producing a hepatotoxin 'microcystin' which in high concentrations could be fatal to humans, finfish, shellfish, birds and pets as well as various other forms of aquatic life. Ingestion of significant levels of the toxin microcyst can cause liver damage and dysfunction in humans and animals. *M. aeruginosa* bloom in Muttukadu backwater can lead to fish mortality due to high ammonia content.

This toxic bloom is of serious threat to environmental conservation, fisheries and loss of biodiversities. These blooms reduce the water transparency and normal dissolved oxygen content of the water column leading mass mortality of fishes. Non-point sources of pollution are considered to be an important factor enhancing algal blooms and disrupt the normal phytoplankton composition in this backwater. The pollution abatement facilities like the sewage and domestic waste treatment at Muttukadu area are inefficient, resulting in indiscriminate discharge of wastes into the backwaters.

2.4.3.1 Introduction

The bloom caused by the blue-green algae, an ecologically and economically important cosmopolitan species, affects the aquatic organisms and is of global concern (Sivonen 1996). The blooms of *Microcystis* typically flourish in warm, turbid and slow-moving waters, and the highest biomass occur in eutrophic waters (waters enriched with nitrogen and phosphorous).The species also requires sufficient light intensity to conduct photosynthesis, which results in blooms. This alga is capable of producing a hepatotoxin 'microcystin' which in high concentrations could be fatal to humans, finfish, shellfish, birds, and pets. The toxin microcystin is capable of damaging liver and dysfunction in animals and humans when ingested in significant concentration, but, however, no deaths from ingestion of microcystins have been reported so far in humans. Yoshinaga et al. (2006) had reported that *Microcystis* growth produces foul odour and unsightly scum, preventing recreational use of water bodies, hampering the treatment of water for drinking and clogging irrigation pipe. *M. aeruginosa* thrives both in freshwater and in moderately brackish water, often forming dense blooms in mid- to late summer and fall to the bottom sediments in autumn (John et al. 2002).The bloom is of particular concern in large water bodies and aquaculture ponds where it affects the productivity of zooplankton and thereby reducing fish production (Chellappa et al. 2004). Studies by Federico et al. (2007) and Ghadouani et al. (2006) have demonstrated the effect of *Microcystis* or its toxins on the growth and survival of zooplankton. Legrand et al. (2003) have highlighted that *Microcystis* can affect phytoplankton community composition through allelopathy (the phenomenon has been described in detail in Chap. 3).

Toxin-producing cyanobacteria in lakes and reservoirs form a threat to humans, birds and fishes as well as various other forms of aquatic life. Freshwater systems have become serious water quality problems which also threaten human and animal health (Chorus and Bartram 1999; Carmichael 1992). Nevertheless, a number of scientists have worked upon the occurrence of algae in marine and brackish waters (Thajuddin et al. 2002; Selvakumar and Sundararaman 2007; Reginald 2007; Velankar and Chaugule 2007); this is the first report on the appearance of harmful algal bloom of the colonial form of *Microcystis aeruginosa* in Muttukadu backwater, southeast coast of India, on June 2012.

2.4.3.2 Characteristics of the Bloom

Microcystis aeruginosa mainly occurred as colonial morph, i.e. single cells combined together in groups to form colonies which tend to float near water surface (Fig. 2.8). The cell density of *M. aeruginosa* was recorded as 6×10^8cells l^{-1} which shared 95–98% of the total phytoplankton density during the bloom phase. *M. aeruginosa* populations accounted for significant proportions of total phytoplankton biomass in the surface water column. The concentration of silicate was higher when compared with other nutrients. During *M. aeruginosa* bloom, toxic or oxygen

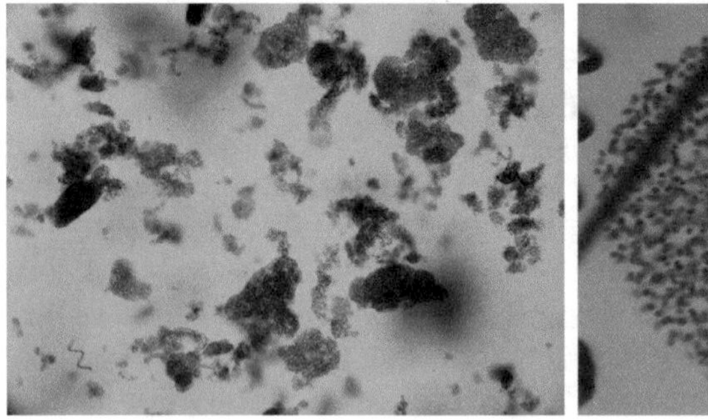
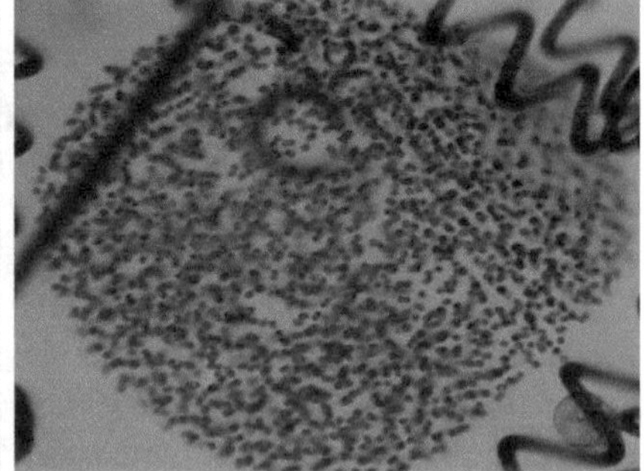

Fig. 2.8 Bloom-forming cyanobacterium *Microcystis aeruginosa* in Muttukadu backwaters, India

deficiency led to mortality of fishes which were washed along the shore. Due to the presence of large concentration of *M. aeruginosa,* the water colour turned green, slimy in nature along with foul smell. Large quantities of effluents from domestic source and effluents from factories located along the bank of Muttukadu estuary are discharged into the backwater. Moreover, pollution sources which enhance eutrophication process of these regions must be obstructed. Necessary preventive steps should be implemented to control pollution sources which accelerate eutrophication process of these regions. Nutrient content in surface waters, receiving high solar radiation, can influence the abundance of phytoplankton. This is sure to increase the phosphorous content of the surface water where this nutrient will otherwise act as a limiting factor. This could be the reason for a bloom of the cyanobacteria *M. aeruginosa* during this period as endorsed by other workers (Yoshinaga et al. 2006).

Murrell and Lores (2004) had stated that cyanobacteria dominance often occurs when water temperature rises above 20 °C which is also seen in subtropical waters, including coastal systems. Salinity has been considered to be another important factor influencing the production of cyanobacteria blooms (Blackburn et al. 1996; Hobson et al. 1999), although many species are also capable of grow and bloom over a wide range of salinities (Reed and Stewart 1988) from freshwater in lakes and rivers, transitional brackish environments, such as estuaries, to oceanic waters and even in hypersaline lakes. This is to be referred that salinity seems to have no impact on bloom formation which reflected the findings of previous researchers (Fay 1983; Falconer et al. 1994; Reynolds et al. 1981).During daytime, microalgal photosynthesis produces excess oxygen than that of highly dense algal biomass respiration, resulting in higher dissolved oxygen concentration (Capone et al. 1997). Water transparency was measured with the help of standard Secchi disc, and it was observed that *Microcystis* bloom reduced water transparency through highly dense algal biomass. Both nitrate and phosphate are prime indicators of water quality in aquatic environment. Nitrate is highly soluble, readily leachable from soil and for plant growth-limiting factor nutrient which when becomes easily available in waterways contribute to harmful algal blooms (HABs). High concentration of phosphate during bloom day might be due to decomposition by microbial process as reported by previous workers (Subba Rao 1969; Raghuprasad and Jayaraman 1954; Sahayak et al. 2005).

Ammonia concentration showed high values coinciding on the day of maximum cell density which might be due to the high demineralization ability of cyanobacteria to produce ammonia through the process of nitrogen fixation (Chang et al. 2000). Due to the presence of high ammonia, toxin (Mirocystin) or oxygen deficiency, *M. aeruginosa* bloom can lead to the mortality of fishes in Muttukadu backwater.

Muttukadu estuary, being a bar built estuary, recorded no inundation of seawater during the bloom due to closing of the bar mouth. It is well known that the stagnant condition resulting lack of flow of water might have resulted 'dystrophic' condition supporting large density of *M. aeruginosa* which occupied around 98% of the biomass. The estuary is of great importance owing to the economic and ecological services it provides to the people inhabiting the region. Regular monitoring of the this backwater is essential for the conservation of biodiversity and human health since *Microcystis aeruginosa* blooms has been reported as highly toxic species. Results of water quality analysis such as water transparency, dissolved oxygen and other physico-chemical parameters revealed that backwaters of Muttukadu were affected by *M. aeruginosa* bloom. These blooms form a thin layer similar to film of oil on the surface of the water, thereby decreasing water transparency and normal oxygen content of the water column and thus resulting in mortality of fishes. Non-point sources of pollution seemed to activate algal blooms and disrupt the normal phytoplankton composition of this backwater. The pollution abatement facilities like the sewage and domestic waste treatment at Muttukadu area are inefficient, resulting in indiscriminate discharge of wastes into the backwaters. Regular monitoring of harmful algal blooms and careful and systematic assessment of the physico-chemical characteristics of the Muttukadu backwaters are essentially required to conserve the fish stocks and safeguard the human beings.

2.4.4 Kerala Coastal Regions, India

An intensive study on the occurrence of bloom caused by marine radiophyte *Chattonella marina* (Subrahmanyan) (Y. Hara and M. Chihara 1982) and its adverse effect on the fisheries was carried out by Jugnu and Kripa (2009) along the Kerala coast (approx. 580 km in length) during October 2001 to September 2003. The climate is chiefly wet and maritime tropical, largely governed by the seasonal heavy rains due to southwest monsoon.

Blooms of *C. marina* algae were observed along the northern coastal part of Kerala, and the maximum cell density reached as high as 28×10^7 cells l^{-1} in September 2002 and 135×10^5 cells l^{-1} in September 2003, leading massive fish kills. Hypoxic conditions or hypoxia (depletion of oxygen) prevailed during bloom phase when the concentration of dissolved oxygen ranged from 0.22 to 1.92 mg $^{-1}$. The fish landing data using gears which have the ability to cover a distance of 55 km from the shore), viz. outboard trawl net (hand trawl) (OBTN), outboard drift net (OBDN), outboard ring seine (OBRS), outboard gill net (OBGN) and country craft gill net(CCGN), were used to ascertain the impact of the bloom on the coastal fishery resources of the region.

The landings of fishes were decreased which belonged to lower trophic levels. The zooplanktivorous fishes like *Stolephorus* spp., *Thryssa* and *Leiognathus* were totally absent, but the predatory fishes (of higher trophic level) such as *Euthynnus*, *Trichiurus*, *Carcharhinus*, *Saurida*, *Scoliodon*, *Scomberomorus*, and *Sepia* spp. were increase in catch. There was significant variation in catch rate between the bloom and the non-bloom period especially for *Cynoglossus* spp., *Johnius* spp., *Thryssa* spp. and *Parapenaeopsis stylifera* caught in the outboard trawl net. The average taxonomic distinctness was lower during the bloom period, and a distinct shift in the community structure was noticed. However, the effect of the *C. marina* bloom on fish community was short-lived, and the species diversity was regained soon when the bloom receded.

2.4.4.1 Introduction

Harmful microalgal blooms cause severe problems to human health, coastal activities and fishery resources throughout the world. A normal diatom-dominated bloom is usually considered favourable for the fishery production of the region, whereas harmful blooms are unfavourable as the fishes tend to avoid such areas of bloom either due to the heavy biomass or due to the production of toxic substances which are harmful to it. Raphidophytes have been recognized to produce neurotoxic, haemolytic and haemoagglutinating compounds as well as superoxide and hydroxyl radicals. These are causative factors for severe damage to fish gills leading to osmoregulatory problems and mortality (Endo et al. 1992; Tanaka et al. 1994; Ahmed et al. 1995). Both the radiophytes, *Chattonella antiqua* and *Chattonella marina,* are the major species related to fish mortality in aquaculture farms in Japan.

Subrahmanyan (1954) first described the species in connection with occurrence of a bloom along the Calicut coast of North Kerala resulting in fish and faunal mortality and fish avoidance along the Indian coast since the nineteenth century (Hornell 1917; Chacko 1942; Jacob and Menon 1948). Blooms caused by *Chattonella marina* persist for comparatively longer period of time in nearshore regions of Calicut coast than in other regions. There was a drastic change in the ecosystem, affecting the fishery resources, and this was reflected in the catches landed in the region. For such studies taxonomic measures are used. A far-reaching change, representative of what is happing to the community as a whole, is indicated by looking at the taxonomic relatedness statistics based only on presence/absence data (Clarke and Warwick 2001). Rogers et al. (1999) investigated on the taxonomic structure of demersal fish assemblages in North Sea, English Channel and Irish Sea applied to assess the impacts of beam trawling. Similarly taxonomic distinctiveness has been applied for demersal fish (Hall and Greenstreet 1998), star fish and brittle stars of polar regions (Piepenburg et al. 1997) as well as for Atlantic starfish assemblages (Price et al. 1999). To estimate the negative impact of the bloom on the coastal fishery resources of the region, the fish landing data of north Kerala region was collected covering following three phases: during bloom and non-bloom period and the specific time period related to the harmful algal bloom (HAB).

2.4.4.2 Plankton Analysis

Regular sampling was undertaken monthly for 2 consecutive years (October 2001–2003) at Chombala, a small Village/hamlet in Kozhikode District, along the north Kerala coast. The sampling frequency was increased during the bloom period, and sampling was carried out on the 1st, 8th and 22nd day till the bloom phase was over. This was confirmed by the total absence of the species from the bloom region. Additional sampling sites such as Konadu and Kappad villages were also considered to get a holistic feature when more area was affected due to occurrence of the bloom.

Phytoplankton samples were collected with the help of a phytoplankton net (mesh size 30 μ and mouth diameter 50 cm) which was hauled horizontally for 15 min from a boat. Phytoplankton sample collected in the bucket at the end of the net was finally stored in a container and preserved in 4% buffered formalin for further analyses. Sedimentation method was adopted for quantitative estimation of phytoplankton. An aliquot of 1 ml of the phytoplankton was taken and carefully counted with the help of a Sedgewick Rafter counting chamber. The process was repeated in triplicates, and the mean value of phytoplankton was considered and expressed in cells l^{-1}. The identification of species into different taxonomic categories was done with the help of taxonomic keys published in several research papers (Subrahmanyan 1968, 1971; Halegraeff et al. 1995; Tomas 1996). Concentration of dissolved oxygen (DO) was analysed using the method of Strickland and Parsons (1972).

2.4.4.3 Impact on Fisheries

For estimation of *Chattonella marina* bloom effect, the fishery data generated by the Fishery Resource and Assessment Division (FRAD) of Central Marine Fisheries Research Institute (CMFRI) were considered. Stratified multistage random sampling method, devised by Alagaraja et al. (1992), was taken into consideration for estimation of the fishery landings. The landing data by the different gears operating along the Calicut coast (Lat 11° 43′ N, Long 75° 33′ E) zone was collected, as because these areas are affected with regular bloom occurrences. The macro-level changes observed in the preliminary database were used to trace the micro-level species-specific variations.

The following inputs were used to identify the main marine commercial fishery species which are affected by *C. marina* blooming. The catch from the fishing crafts operating in the Calicut region are landed in two major harbours, viz. Chombala fishing site and Puthiyappa along with 3 major landing centres and 11 minor landing centres.

While assessing the bloom impact, the variation in magnitude of total monthly landings during October 2001 and December 2003 was considered. The alga reached the peak density of 28×10^7 cells l^{-1} during September 2002 (Bloom I), and subsequently it reached a bloom density of 17×10^4 cells l^{-1} and was taken as Bloom II in September 2003.

The catch per unit effort per day (CPUE) for these species was recorded to compare the immediate pre-bloom and post-bloom period with that of bloom period. Thus a micro-level analysis was done exclusively for 6 months as follows:

1. Pre-bloom: May, July (excluding the trawl ban period) and August were considered as pre-bloom period since during June and July the fishery was low due to inclement weather conditions. In addition, the trawl ban enforced by the Kerala government was also imposed during this period.
2. Post-bloom period: Both October and November when *Chattonella marina* was not present at the site and also when there was no bloom-causing species.

2.4.4.4 Effect on Community Structure: Taxonomic Diversity Indices

Species richness is largely dependent on sampling effort, which is highly variable in this case, and since the exclusively data on fish weight were available and not the numbers, the taxonomic diversity indices, viz. average taxonomic distinctness, were calculated (Clarke and Warwick 1998). Taxonomic diversity indices are a type of diversity indices which consider the taxonomic relation between different organisms in a community. This reflects the average taxonomic distance between any two organisms, chosen at random from a sample, and this was based on the assumption that a sample containing species belonging to distantly related species is more diverse than a sample containing closely related species. This is usually defined from a Linnaean classification and requires an aggregation file in addition to the data worksheet. The aggregation file is used as a look up table which gives the taxonomic relationship and the distance apart of any two species in the sample. The taxonomic tree was made considering of seven taxonomic levels starting from the genus level and extending to family, suborder, order and class up to the level of phylum. All branch lengths were assigned equal weights.

During September 2002 and September 2003, the harmful alga *Chattonella marina* (Subrahmanyan) Y. Hara & M. Chihara was present in the algal community exclusively in the northern Kerala region. An intensive and high-density bloom of *Chattonella marina* occurred during the first week of September 2002, exhibiting a discontinuous distribution and stretched over a distance of approx. 50 km along the coast. The distribution was ascertained by estimating the plankton densities at various points along the coast within this region. The muddy green-coloured bloom was very pronounced at Kappad, where it extended from the shoreline to about a distance of ~3 km towards the sea. It was visible as streaks and patches in other region.

2.4.4.5 Phytoplankton Characteristics of the Harmful Algal Bloom

On the first day of the bloom, *C. marina* reached the peak density of 28×10^7 cells l^{-1} at Kappad along with other microalgae present in high densities as follows: *Coscinodiscus asteromphalus* Ehrenberg at a density of 8×10^4 cells l^{-1}, *Pleurosigma normanii* Ralfs at 2×10^4 cells l^{-1} and *Noctiluca sigma* at a density of 4×10^4 cells l^{-1}. Density of *C. marina* dropped considerably to 4234 cells l^{-1} on the 8th day. The other phytoplankton species were also present in considerably low numbers as follows: *C. asteromphalus* (1312 cells l^{-1}), *P. normani* (22 cells l^{-1}) and *N. sigma* (552 cells l^{-1}).

C. marina bloom was recorded in the second week at this site, being present in high density of 1.7×10^4 cells l^{-1}. The harmful alga *Noctiluca scintillans* was also present at a density of 100 cells l^{-1} along with *C. asteromphalus* at a density of 24,700 cells l^{-1}, *P. normani* at a density of 2300 cells l^{-1} and *N. sigma* at a density of 8100 cells l^{-1}. On the 14th day, the concentration of *N. scintillans* reduced to 2990 cells l^{-1} to 60 cells l^{-1} and that of *C. asteromphalus* to 2800 cellsl^{-1} at the site.

A total of 66 genera of finfishes (33 pelagic and 33 demersal), 13 crustaceans and 4 cephalopods were recorded by the FRA Division for the region. Among the crustaceans, mantis shrimps, crabs and lobsters were present while squids, cuttlefishes and octopuses formed the major molluscan components. Based on the presence or absence of the species a generalized impact grading of species was done and the variation in the magnitude of the monthly landing. Accordingly, the species were graded into four categories as follows:

1. Species present exclusively during the bloom
2. Species present in the catch throughout the year, except during the bloom period
3. Species present in the catch throughout the year including the bloom period but whose magnitude of landings decreased during the bloom period compared to the non-bloom period
4. Species which were present in the catch throughout the year including the bloom period but whose magnitude of landings increased during the bloom period in comparison to non-bloom period

The general impact on fishery was mass finfish and shellfish (bivalve molluscs) mortality in the region between Puthiyappa and Kappad. Fishes were killed and were thoroughly washed all along the shore between these two stations. Mortality was recorded mainly in demersal fishes; a major percentage of the dead fish were eels (Anguillidae) followed by sciaenids and croakers. In addition, mortalities were also observed in other fishes, viz. *Epinephelus* sp., *Otolithes* sp., *Cynoglossus* sp. and *Johnius* sp. The fish and bivalve samples analysed at Central Institute of Fisheries Technology did not show any toxin in the body flesh. The fishery of the region was severely affected during the occurrence of red tide.

A substantial algal biomass results to a rich zooplankton crop and high survival of young fish, mainly larvae and juveniles, as endorsed by Raymont (1980) and Shumway (1990). Fluctuations in the phytoplankton standing stock in the region therefore reflect the variability in the recruitment and survival of edible fish stock. Analyses of fish landings of the Calicut region indicated that the coastal fishery of the region is affected by the *C. marina* bloom; nevertheless it lasted for short time span.

Fish mortality along the Indian coast (~7517 km) was coincided with the occurrence of bloom by phytoplankton genera, e.g. *Noctiluca*, *Trichodesmium* and *Gymnodinium mikimotoi.* The avoidance of the bloom region by fishes along the Calicut coast was mainly due to their high power of mobility. Bhimachar and George (1950) observed that commercially important shoaling species like sardine and mackerel shift from foul water mass to comparative pristine region in adjacent coastal region. In these waters this shifting of the fishery has been observed during the excessive production of *Noctiluca* sp., *C. marina* with similar avoidance of the bloom region of this alga by the fishes (Subrahmanyan 1954), *Nitzschia* and *Oscillatoria.* The euglenoid was revealed that the oil sardine and mackerel which were landed in substantial quantities during the months preceding *Trichodesmium* spp. bloom subsequently declined with the bloom and revived with the subsidence of the bloom (Prabhu et al. 1971), whereas trawl operations in the areas of dense bloom along Goa coast showed fish catches similar in size and composition as in non-bloom areas (Devassy et al. 1978). A dense bloom of *Trichodesmium erythraeum* was found to severely affect the fisheries of the Minicoy Island, the archipelago, southern most atoll of Lakshadweep with the fishes completely avoiding the area during the bloom period (Nagabhushanam 1967). During *Noctiluca* bloom fall, due to the avoidance of the bloom area by the fishes, has also been reported (Devassy and Nair 1987; Shetty et al. 1988). Mass mortality to the marine benthic fishes along the west coast due to *G. mikimotoi* bloom has been reported (Karunasagar and Karunasagar 1992).

Development of fishing craft and gear resulted in a significant change in the fishing activity of Calicut. Seasonal fishery in the inshore waters of the Calicut coast has been chiefly related to plankton production (Chidambaram and Menon 1945; Subrahmnayan 1959). Considering the landings of the region, the catch had reduced subsequently during the bloom period in both the years. Bloom was a prolonged one and resulted in large-scale mortality of finfishes and shellfishes. Majority of the demersal fishes like eels, groupers, sciaenids and croakers were affected. Fishes are reported to be killed by anoxia during exposure to C. *marina* red tides (Matsusato and Kobayashi 1974; Ishimatsu et al. 1990). Neurotoxins have been isolated from these organisms, and it is observed that the neurotoxin fraction is more toxic to fishes than the hemolytic and hemo-agglutinating fractions (Onoue and Nozawa 1989). The toxin analysis at Central Institute of Fisheries Technology was positive only for the water samples but not for any of the fish and faunal samples from the region indicating that toxin does not accumulate and cause any negative effects as the toxin analysed was not positive for fish. Bloom I had resulted in the cancellation of all inshore fishing operations in the bloom area about 3 weeks with a severe negative impact on the fishing economy of the region. The mechanized trawl net operations were cancelled completely. There was an absence of fish shoals starting from the pre-bloom and extending up to the post-bloom periods. The alga has a benthic cyst in its life cycle, and its germination during the pre-bloom month could have altered the water quality as evidenced by the low dissolved oxygen (DO) values, high total suspended solids (TSS) and toxin production which were unfavourable to the fish and several marine fauna. The presence of toxin in the water which was produced by the alga must have also caused the avoidance of the region by fish shoals.

The community structure change with an absolute dominance of fish community of higher trophic levels was more pronounced, as evidence from AvTD tests (the biological index average taxonomic distinctness). The taxonomic distinctness decreased suggesting a stress and a community structure change was probably linked to the bloom. Clarke and Warwick (1998) identified 14 species from a range of impacted as well as undisturbed UK areas and observed that the Av TD clearly varied in the impacted areas whereas comparatively pristine locations Av TD was similar to the master species list. Based on an impact study of beam trawling in taxonomic structure of demersal fish assemblages in the North Sea, English channel and Irish sea, Rogers et al. (1999) asserted that delta index was distinctly reduced in some areas due to the stress caused by trawling.

The landings of all fishes which belonged to lower trophic levels revealed a decrease in catch. During the blooming of *C. marina* in September 2002, the landing of fishes belonging to lower trophic levels, there was an increase in the catch of sardines mainly from ring seines which may be attributed to the favourable diatom blooms, *Coscinodiscus asteromphalus* Ehrenberg, 1844, which occurred preceding the HABs. An unusually high landing of sardines was also recorded in October 2003, and this coincided again with a bloom caused by the same diatom in the region. There was however a decrease in sardine landings from gears like hand and mechanized trawl nets which generally operate very close to the shore. This might happen because these fishes must have avoided the bloom region due to the irritant property of this alga. Again, since these gears could not venture in these far off regions for fishing due to their restricted mobility, there was a decrease in the catch rate of these fishes in these gears. The decrease in landings in the gears operated near to the shore and the unusual catches of some fishes (e.g. sardines) in outboard drift net distinctly indicate that there exists a tendency to avoid the bloom affected by fishes. The zooplanktivorous fishes (such as *Stolephorus* spp., *Thryssa* and *Leiognathus*) were totally absent, which can be attributed either to the absence of their food from the region or to the presence of toxins which might have been transferred through the food web. The planktonic herbivores can accumulate algal toxins and retain them to a certain extent (White 1981). This might have resulted in the decrease in catch of the groups which are mainly fed on these zooplankton feeders, such as *Caranx* (tropical to subtropical marine fish) and *Johnius* spp. (bearded croaker).

Increase in catch of the predatory fishes belonging to higher trophic level mainly *Euthynnus*, *Trichiurus*, *Carcharhinus*, *Saurida*, *Scoliodon*, *Scomberomorus* and *Sepia* spp. was recorded. During the bloom, these species were present in high levels. During the bloom, these species were present in high quantity occurring during the pre- or post-monsoon period. Except *Sepia* and *Scoliodon*, all the other species were present during bloom II too. The catch of the sharks, *Carcharhinus* sp. and *Scoliodon* sp., had increased considerably during the bloom period. They feed mainly on pelagic and shoaling teleosts like sardine, scad, mackerel, squids, etc. which were present in good density due to bloom. The neurotoxins of *C. marina* were more toxic than other cytotoxins produced by this species (Onoue and Nozawa 1989). Exposure of fish to *C. marina*, the red tide-causing species, has been found to result in asphyxiation and erratic swimming behaviour in fishes (Endo et al. 1992).

Cell density also plays a major role in detecting the presence of this predator species. With increase in cell density and surface accumulation of the bloom, total avoidance of the bloom area by these fishes was marked. In the second year the densities were however lower and the bloom sustained for a short period of time. Hence the catch of these species were high in the bloom II in September 2003. The catch of all shrimps had decreased during the bloom period due to depletion of the food-comparing species of phytoplankton *Fragilaria*, *Coscinodiscus*, *Pleurosigma*, *Navicula*, *Cyclotella*, etc. During the bloom of the dinoflagellate *Gyrodinium aureo-*

lum, a similar decrease in shrimp fishing was reported which revived after the dinoflagellates dispersed (Tangen 1977).

The results reveal that the fishery of the Kerala seems to be adversely affected by the *C. marina* bloom. However, the negative impact of the bloom on macrofauna was short-lived increasing immediately after the subsidence of the bloom. According to Gosselin et al. (1989), the kills of adult fishes are sporadic events with limited impacts on fisheries. The emergence of fish larvae and post larvae at a time when the planktonic food web is contaminated by algal toxin could lead to a significant reduction of early survival and threaten recruitment to local stocks. Hence the effect of the algal toxins on the different trophic levels has to be studied in detail.

Madhu et al. (2011) reported intense brownish discolouration in the surface waters of Cochin barmouth on October 21, 2009, which found to extend a few kilometres in the coastal waters. Continuous monitoring was carried out in the region for a period of 5 days (October 21–25, 2009). Substantial data were generated by continuous monitoring in the region pertaining to changes in water quality and phytoplankton community associated with the bloom. Relatively high level of inorganic nutrients (ammonia 0.5–9.7 μm, nitrate 1.3–2.9 μm, phosphate 1–2.3 μm, silicate 23.5–39.9 μm) and chlorophyll *a* (av. 56.8 ± 23.7 mg m^{-3}) concentration were recorded during the bloom period. Microscopic analyses revealed that the discoloration was caused by an unarmoured toxic dinoflagellate, *Karenia mikimotoi* Miyake & Kominami ex Oda (Karenia, Gymnodiniaceae) belongs to Gymnodiniales, which causes massive fish mortality by releasing hemolytic compounds.

Very high chlorophyll *a* concentration (av. 56.8 ± 23.7 mgm^{-3}) was observed with a maximum of 85.8 mgm^{-3} on October 24. The phytoplankton density varied from 336.1 × 10^4 cells l^{-1} to 1568.5 × 10^4 cells l^{-1}, with a maximum density on October 21. Among these, dinoflagellates were numerically dominant group (72.2–99.2%) followed by diatoms (0.8–27.8%). Microscopic analyses revealed that the discolouration was caused by an unarmoured dinoflagellate, *Karenia mikimotoi* Miyake & Kominami ex Oda 1935 (*Karenia*, Gymnodiniaceae, Gymnodiniales), mainly existing in the temperate coastal environment (Zhang et al. 2009) and formerly known as *Gymnodinium mikimotoi* (Fig. 2.9). Cells appeared oval in shape and dorsoventrally compressed with 20–30 μm long and 18–25 μm wide. Epitheca seemed to be broadly rounded and smaller than the notched, slightly bilobed hypotheca. Nucleus was not clearly visible as it was surrounded by many shapeless yellowish-brown coloured chloroplasts. Other morphological characteristic of this species matched completely with the description of Tomas (1997).

The density of *K. mikimotoi* in the bloom period ranged from 701 × 10^4 cells L^{-1} to 1550 × 10^4 cells L^{-1} which constituted 70–99% of the phytoplankton assemblage. Maximum abundance of *K. mikimotoi* was encountered on 21, whereas on October 25, the density of *K. mikimotoi* is reduced to 0.4 × 10^4 cells L^{-1}. The bloom also contained other dinoflagellates (*Prorocentrum micans*, *P. gracile*, *Dinophysis acuta*, *Ceratium furca*) and diatoms (*Skeletonema costatum*, *Nitzschia longissima*, *N. sigma*, *Coscinodiscus radiatus*, *Pleurosigma normanii*) in moderate levels. Abundance of *Skeletonema costatum* (11.75 × 10^4 cells L^{-1} to 326.5 × 10^4 cells L^{-1}) was associated with *K. mikimotoi* bloom throughout the observation. *K. mikimotoi*

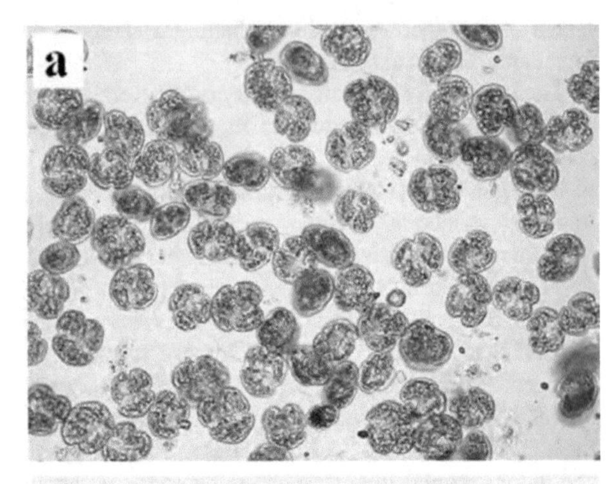

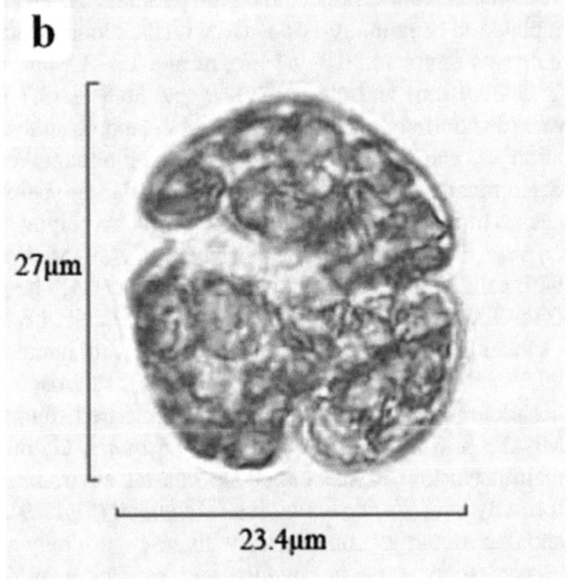

Fig. 2.9 Microscopic view of *Karenia mikimotoi* under inverted stereoscope binocular microscope (OLYMPUS) at magnification of (**a**) 100 × and (**b**) 400×. (**b**) Enlarged view (400× magnification) of *Karenia mikimotoi* under inverted stereoscope binocular microscope (OLYMPUS)

is an autotrophic red tide dinoflagellate and has been linked to massive fish kills including fin fishes as well as benthic fauna of the temperate coastal waters worldwide (Zhang et al. 2009; Takayama and Adachi 1984; Ulrich et al. 2010). This bloom was first recorded in Irish waters, occurring regularly and leading to major kills of marine organisms including fish, shellfish, brittle stars and farmed salmon and cod (Davidson et al. 2009). In China, bloom of *K. mikimotoi* causes massive fish kills in 1998, resulting economic losses estimated at US$ 40 million (Lu and Hodgkiss 2004). Toxic blooms of *K. mikimotoi* have become increasingly frequent worldwide, and this bloom can produce toxic glycolipids along with numerous other unidentified compounds (Ulrich et al. 2010). Even though *K. mikimotoi* is ichthyotoxic and hemolysins have been identified as exotoxins, the exact mechanism of toxic effect to marine organisms remains unclear (Zhang et al. 2009).

Sporadic events of brownish discolouration and associated fish kills have frequently been reported from the Cochin inlets and coastal waters (Naqvi et al. 1998; Sahayak et al. 2005). Nevertheless, there was no fish kills visible from the area during the occurrence of present bloom, probably because the bloom-affected waters might be diluted with river water and shifted to the adjacent coastal waters by tidal effect. Even though many reports are available on red tides, the information on the incidence of *K. mikimotoi* bloom in the Indian of *K. mikimotoi* in Indian waters proves the theory of aggravation of harmful algal bloom (HAB) events by ballast waters and ocean currents (Smayda 2007). Since both the Cochin and Mangalore ports are the busiest ports in the west coast of India, the amount of disposal of ballast water is bound to increase, which can introduce alien micro- as well as macro-organisms and disrupt the native food chain. Reports on HABs worldwide have been proved that ballast water is a significant transport vector for global dispersal of toxic microalgae (Hallegraeff and Gollasch 2006). Therefore, appearance of *K. mikimotoi* in the Cochin barmouth may be an indication of a shift in the phytoplankton community as a result of increased anthropogenic activity in the Indian port, which required further studies.

2.4.5 Gulf of Kachchh, Gujarat, India

Gulf of Kachchh Marine National Park (MNP), comprised of 42 islands, is one of the marine biodiversity rich habitats. It is also known as non-divers' paradise where a plethora of marine organisms can be viewed during low tide (Nair 2013; Adhavan et al. 2014a). This marine protected area is surrounded by multifarious industries such as petrochemicals, fertilizer, shipbuilding, thermal power, salt works and several other small-scale industries that drain their effluents into the sea (Adhavan et al. 2014b). In addition, the southwest monsoon downpours the region and washes the nutrients along with other pollutants from land to sea. It stimulates the nitrification in the Gulf of Kachchh waters and further results in algal bloom every year from October to February. The stimulating factors such as anthropogenic activities, hydrographic changes and impact of climate change are mainly responsible for abrupt growth of algae that causes intensive effect on the coastal ecosystems. The bloom had adverse impact on the mangrove ecosystem at Narara coastal regions and settled over the dead portion of the massive form of hard corals at Pirotan Islands. The seaweed bloom indicates the decrease of fish grazers in the coast. Mortalities were also reported for fish and other animals including mullet, mudskippers and crabs due to occurrence of algal bloom in the mangrove areas. Further, fishes were found avoiding the bloom-affected areas that negatively affects the livelihoods of local communities who rely on traditional fisheries in the region.

Algal blooms occur as a natural phenomenon when environmental conditions promote the rapid growth of algae, which is experienced by the majority of coastal countries. A severity of such algal blooms depends upon the nutrient enrichment level especially phosphorous and nitrogen contents in the ecosystem. During regular monthly survey for biodiversity assessment under the Integrated Coastal Zone Management Project (ICZMP) (2013–2015), algal enrichment on intertidal reef environment of Pirotan Island, Narara and Poshitra reef was noticed as well as in the mangrove ecosystem at Narara coastal area. *Avicennia marina* formed the dominant species followed by sparing distribution of *Ceriops tagal* and *Rhizophora mucronata*. The algal species *Ulva lactuca* (family Chlorophyceae) was found to be dominant in every season during algal blooms, followed by *Sargassum cinereum* and *S. tenerrimum* (family Phaeophyceae). Other associated species such as *Ulva prolifera*, *Caulerpa racemosa*, *C. taxifolia*, *Ectocarpus siliculosus* and *Padina tetrastromatica* were also present, but their abundances were comparatively less.

The bloom drastically affected the mangrove ecosystem at Narara coastal regions and settled over the dead portion of the massive form of hard corals at Pirotan Islands. It was also observed that algal mat covered the pneumatophores (breathing roots) and leaves of mangroves which hampers breathing and photosynthetic processes. They formed a thick coat over the sedentary organisms such as corals, sponges and anemones and restricted the sunlight which might have affected the primary productivities of their symbionts. The seaweed bloom indicates the decrease of fish grazers in the coast.

During this algal bloom, fishes along with other aquatic animals, viz. mullet, mudskippers and crabs, were also found dead in the mangrove regions. Moreover, fishes were able to escape the bloom-affected regions, and such phenomenon had largely affected the livelihoods of local communities who rely on traditional fisheries in the region. The seaweed bloom is considered as a potential biological indicator for pollution and sharp fall in fish landings in the coast.

It is worth to mention that the ecosystems would be grossly affected if the situation continues with a decrease in fish population which would also affect the recruitment process of the sedentary organisms. Manual harvesting of seaweed is strongly recommended as this has multiple uses, such as manure production, pharmaceuticals, cosmetics industries, etc. This should be operated outside the Marine National Park without affecting the ecosystem which would serve as an ecofriendly alternate livelihood for the fisher folk along the Gulf coast.

2.4.6 South Andaman Island, India

The Andaman and Nicobar Islands, India, comprise pristine and diverse terrestrial as well as marine ecosystems. Diverse mangrove and coral reef ecosystem supports unique marine flora and fauna. However, increasing human encroachment continuously destructing the mangrove vegetation, coral reefs and seagrass beds along the South Andaman Islets disrupt the ecological balance. Anthropogenic pollutants of industrial, domestic and agricultural origin are deteriorating coastal water quality of this previously known pristine island region (Sahu et al. 2013).

An intense *Trichodesmium erythraeum* bloom is reported in Burmanallah coast of the region of Port Blair in South Andaman Island on April 10 2013 (Karthik and

Padmavati 2017). The discolouration of the surface water noticed varied from pale brown to pinkish red. Microscopic examination of the surface water revealed the blooming of *T. erythraeum*, and its density was reported to be around 43,000 Cells. ml^{-1}. The elevated nutrient concentrations trigger bloom incidences of *Noctiluca scintillans*, *Trichodesmium* spp. and *Chaetoceros curvisetus* which are continuously increasing along the densely populated bay regions (Eashwar et al. 2001; Dharani et al. 2004; Sahu et al. 2014a, b; Begum et al. 2015). Burmanallah coast is an extremely wave-affected region found in the Port Blair, South Andaman Island, India. The entire area is bay shaped, with freshwater influxes on both the ends. All these influxes are bordered by dense mangrove forests. The anthropogenic influence is quite low here to the other coastal waters of South Andaman Island.

Trichodesmium sp., a marine nitrogen-fixing cyanobacterium, forms extensive surface blooms discolouring vast regions of tropical and subtropical seas. It is one of the common bloom-forming species found in tropical to subtropical waters particularly in the eastern tropical and subtropical Pacific and Arabian Sea contributing >30% of algal blooms of the world (Westberry and Siegel 2006). *Trichodesmium* sp. bloom produces many harmful effects, sometimes causing damage to coastal fish and shellfish fauna (Bhat and Verlencar 2006). Followed by frequent occurrence of *Trichodesmium* sp., bloom in Indian waters, it has been reported more frequently in the west coast (Sarangi et al. 2004; Rajagopalan 2007) as compared to east coast (Jyothibabu et al. 2003; Satpathy et al. 2007). Generally, the bloom of this filamentous alga occurs during hot weather with brilliant sunlight and stable high salinity (Arun Kumar et al. 2012). Occurrence of algal bloom caused by *Trichodesmium erythraeum* which occurred in the coastal waters of Burmanallah, South Andaman Islands, on April 2013 has been described.

A total of 77 species and 39 genera were identified in the Burmanallah coastal waters, comprising of the following dominant taxonomic groups: diatoms (56 species, belonging to 28 genera) and dinoflagellates (18 species and 9 genera). Silicoflagellate comprised two species belonging to one genus, and the bloom species included Cyanophyceae *T. erythraeum*. However, the event led to the exclusion of other phytoplankton species, but no fish mortality was recorded. Nevertheless, some phytoplankton species still persisted in small numbers, regardless of bloom intensity. Certain genera such as diatoms *Bacteriastrum* sp., *Biddulphia* sp., *Chaetoceros* sp. and *Coscinodiscus* sp. and dinoflagellates such as *Ceratium* sp., *Cochlodinium* sp., *Dinophysis* sp., *Gonyaulax* sp., *Gymnodinium* sp., *Lingulodinium* sp., *Oxytoxum* sp., *Prorocentrum* sp. and *Protoperidinium* sp. were recorded. *T. erythraeum* resulted in an increase in temperature during the bloom. Most of the marine blue-green algae exhibit substantial growth in the temperature ranged between 25 and 35 °C (Rajagopalan 2007). Earlier reports (Arun Kumar et al. 2012, 2015; Karthik et al. 2012; Narayana et al. 2014; Sahu et al. 2014a, b) also showed the prevalence maximum temperature during the *T. erythraeum* bloom period (32–34 °C). Previous studies (D'Silva et al. 2012; Sachithanandam et al. 2013) reported around 22 algal blooms along the west (10 algal blooms) and east coasts (12 algal blooms) of Indian Ocean including Andaman Sea from 1908 to 2013. Moreover, majority of algal blooms were dominated by Cyanophyceae in the east coast of

India. The present bloom was noticed during relatively high-temperature conditions 34 °C. Temperature has long been recognized as an important factor that controls *Trichodesmium* abundance. Cyanobacteria are the sensitive to lowest temperatures, and they are apparently excluded in the winter months due to lower temperature (Reynolds 2006). This bloom was occurred during the summer month, which is a dry period or summer season for Andaman and Nicobar Islands. This is contrary to the frequent diatom blooms observed earlier by the authors in the coastal waters of south Andaman which occurred exclusively during the rainy months (Karthik et al. 2014). This once again proved that cyanobacterial species are not dependent on the nutrient flux which is brought by the rainfall unlike the diatoms which are nutrient dependent. A significant reduction in nitrate concentration was noticed during the bloom period. A similar reduction of nitrate concentration during *T. erythraeum* bloom has also observed in the west coast of India (Jyothibabu et al. 2003; Satpathy et al. 2007). In the case of phosphate, it constitutes the most inorganic nutrient which can limit the phytoplankton production in tropical coastal marine ecosystem and there by the overall ecological processes. Many authors have also documented similar increase of phosphate content during the occurrence of bloom of *Trichodesmium* (Madhu et al. 2006; Santhanam et al. 2013); the salinity was found to be the highest (33 PSU) during the bloom period. The salinity condition close to the typical value of 33 PSU and above is known to support the growth and abundance of *T. erythraeum* (Rajagopalan 2007). Previous studies (Mohanty et al. 2010) have also confirmed the fact that stable salinity conditions close to a typical value of 32 PSU and also dissolved oxygen was found to be lower during the bloom period, and above are known to support the growth and abundance *Trichodesmium* sp. Similarly, increase of DO content during *T. erythraeum* bloom has also been reported earlier (Satpathy et al. 2007). This is probably due to the decaying of the cells of the bloom-forming species during sampling. The nitrate concentration was at its lowest during the bloom period, and this is in accordance with the previous studies (Jyothibabu et al. 2003).

T. erythraeum could be also responsible for enrichment of chl a concentrations in a local scale which is an important criterion for fertility of the coastal zone. Occurrence of the present *T. erythraeum* bloom was studied continuously for 1 week period during April 2013 in an enclosed bay where the anthropogenic activities are more due to more nutrient influx into the bay by land runoff. Maximum chlorophyll *a* concentration (0.161 μg. l^{-1}) was also noted by Arun Kumar et al. (2012) in a separate study during the *T. erythraeum* bloom period. Primary productivity is the main criterion for assessing the relative fertility of a particular region.

2.4.7 Arabian Sea and Bay of Bengal

The Arabian Sea lies on the northwest side of the Indian Ocean is unique among low-latitude sea by terminating at a latitude of 25° N and being under marked continental influence. It is the region of monsoon, and biological production is affected by the physical processes altering the vertical flux of nutrients in the mixed layer. Satellite-based observations on ocean colour provide a tool to monitor the biological productivity in terms of phytoplankton concentration. Inter-annual variability of phytoplankton blooms in the northern Arabian Sea during winter monsoon period was investigated by Sarangi et al. (2005). The inter-annual phytoplankton bloom has been studied, and it was most intense during the year 2001 from the point of view of consistent higher chlorophyll concentration around 1 mg/m^3 in the bloom water. The bloom condition has been significantly linked with the change in the physical condition of water, the sea surface temperature gradient, which was about 3 °C rise in sea surface temperature in the bloom and non-bloom condition. Winter planktonic non-toxic bloom (*Noctiluca miliaris*) was investigated by Dwivedi et al. (2012) from Northern western Arabian Sea. A pattern of zooplankton showed unusually high growth in the bloom waters. Fish catch data (% hooking rates of tuna) were obtained from Fishery Survey of India and were used to study response of fish to prevailing high productivity in the bloom waters. Fishing in oceanic waters within Indian EEZ indicated no adverse effect of bloom; rather remarkably higher catches resulted from long-line operations in the bloom waters. *N. miliaris* red tides are commonly encountered in coastal and nearshore regions of the world resulting in a pinkish red or orange discolouration of the water. Active bloom area lies beyond the Indian EEZ where average hooking rate (HR) of tunas is 1.8% from the moderate bloom waters, which is significantly higher than the average value 0.23% from non-bloom waters.

Noctiluca miliaris also referred to as *Noctiluca scintillans* is a marine planktonic dinoflagellate species. *Noctiluca* cells are unarmoured, large (varies from 200 to 2000 microns in diameter) and spherical in shape with a single flagellum (as shown in Fig. 2.10). They are intensely bioluminescent, cosmopolitan species with worldwide distribution in tropical and temperate waters. Colourless cytoplasm and absence of chloroplasts makes this species a non-photosynthetic heterotroph. *Noctiluca* is a phagotrophic and feeds upon other phytoplankton mainly diatoms (Kofoid and Sweazy 1921). *Noctiluca miliaris* is strongly buoyant and bloom-forming species. *N. miliaris* red tides are commonly encountered in coastal and

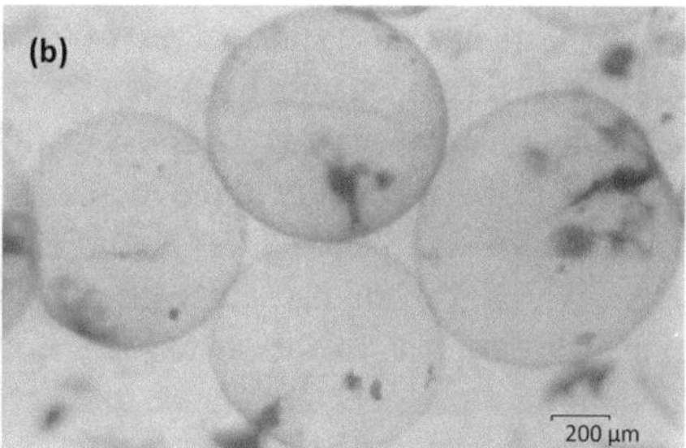

Fig. 2.10 (**a**) *Noctiluca scintillans* red tide in the south Orissa coastal waters, India, on April 5, 2005, and (**b**) individual colony of *N. scintillans*. (Courtesy: K.K. Satpathy, Environment Division, IGCAR)

nearshore regions of the world resulting in a pinkish red or orange discolouration of the water. Though the *Noctiluca* cells are colourless and non-photosynthetic on its own, the pink or red colour is attributed to the ingested material by *Noctiluca* depending on the geographic location (Elbrächter and Qi 1998). In tropical oceans, (Arabian Sea, Indonesia, Malaysia, Thailand, Manila Bay) *Noctiluca* cells harbour green motile flagellates. The presence of these endosymbionts, *Pedionomonas noctilucae* belonging to family prasinophyceae, gives *Noctiluca* blooms green or lime green colour (Sweeney 1976). It influences vast area of the basin for at least 3 months, and, therefore, study of its influence on living marine organisms was considered important. Increase in productivity in the NWAS waters is found to be the highest during the bloom and occurs as early as in December/January. Resulting increased load of particulate organic matter could cause massive sedimentation where bacteria might decompose it consuming dissolved oxygen from the surrounding waters. It would create high demand for oxygen particularly during decay phase in March. Corresponding to higher production on the northwestern side, respiration is also expected to be more aggravating anoxic condition. In an environment like this, fish would be deprived of oxygen. Relatively more intense convection in the NWAS could be another factor for the observed depletion of oxygen. Fish mortality due to bloom-forming dinoflagellate *Noctiluca Scintillans* was recorded in the Gulf of Oman (Al-Azri et al. 2007). Mortality of fish is attributed to the rapid increase in the population of *Noctiluca* and *Trichodesmium* during the months of June, January and October. Over all, fishing operations within bloom waters yielded consistently high tuna hooking rates. Catches declined when fishing was performed in the waters away from influence of the bloom. In addition to response of fish, the zooplankton communities, primary consumers dominated by copepods, were found in abundance during the bloom period. These observations indicate that the bloom is not harmful or repellent for tunas rather it is supportive. Drop in fish catch in 2009 may be attributed to decline oxygen level due to 2009 anomaly.

Two oceanic blooms of *Trichodesmium erythraeum* (Cyanophyta) were recorded in the open waters of Bay of Bengal during 2001 (Jyothibabu et al. 2003). The intense blooms appeared as a thick layer of saw dust on the surface waters. High-integrated primary production (Bloom 1- 2160 mg C $m^{-2}d^{-1}$, Bloom 2- 1740 mg C $m^{-2}d^{-1}$) was obtained in these regions, which indicated the enhancement of primary production in the earlier stages of the bloom. Very low NO_3-N concentrations, undisturbed patches with brownish yellow colour and increased primary production indicated that the blooms observed were in the growth phase. The enhanced primary production could be due to the increased carbon requirement/uptake during the initial phase or the peak of the bloom for constituting the necessary biomass of the algae which, in turn, is required for N_2 fixation. In the bloom-affected regions, zooplankton biomass in the mixed layer was comparatively low. Copepods showed maximum numerical abundance and percentage contribution among zooplanktonic organisms followed by Chaetognaths (commonly known as arrow worm, an exclusive carnivorous marine holoplankton).

Sahu et al. (2015) investigated the changes in mesozooplankton community structure due to occurrence of *Trichodesmium erythraeum* bloom in the coastal waters of south-western Bay of Bengal. Maximum values of zooplankton density (individuals per 10 m^3) reached during pre-bloom period (5.5×10^5) followed by bloom (4.9×10^5) and post-bloom period (4.3×10^5). During pre-bloom period, the copepods shared over 60% of the zooplankton abundance, dominated by the calanoid subgroup (>40%). Carnivore copepods (25%) (Cyclopoids and Poicilostomatoids) were dominant over herbivore (23%) (Paracalanidae) during peak bloom. Release of toxins associated with depletion of oxygen by the decaying biomass could have a negative effect. In contrast, boost in primary production could enhance the fishery potential of the region temporarily. The reduction of copepod density during bloom and post-bloom period might be attributed to the neurotoxic behaviour of *Trichodesmium* or depletion of oxygen due to decomposition of algal stock.

In Indian waters *Trichodesmium* blooms and associated 'red tide' phenomenon were addressed by many researchers in the neritic areas of the west and east coasts of India (Adhikary and Sahu 1992; Devassy et al. 1978; Santhanam et al. 1994a, b; Satpathy and Nair 1996; Nair et al. 1981). However, no reports are available on *Trichodesmium erythraeum* blooms from the open waters of Bay of Bengal. The work provides information on an oligotrophic system that at times becomes highly productive due to *Trichodesmium erythraeum* blooms. In connection with Marine Research-Living Resource (MR-LR) Assessment Programme, observations were made during cruise-192 of *FORV Sagar Sampada* from April 3 to April 30, 2001. Primary productivity studies using ^{14}C technique showed increased integrated primary production in the regions of algal blooms, which were much higher than the average primary production of the Bay of Bengal (225 mgC $m^{-2}d^{-1}$) during this season. Very low NO_3-N concentrations, undisturbed patches with brownish yellow colour and increased primary production indicated that the blooms observed were in the growth phase. Some of the earlier researchers also reported similar situations. Devassy et al. (1978) observed many fold increase in surface primary production in the bloom (500 mgC m^3 h^{-1}) against non-bloom areas (150 mgC m^3 h^{-1}) in the near-shore waters of Goa, while Santhanam et al. (1994a, b) reported around five times increase in surface primary production during the early stages of the bloom (535 mgC m^3d^{-1}) compared to the post-bloom period (150 mgC m^3d^{-1}). The enhanced primary production could be due to the increased carbon requirement/uptake during the initial phase or the peak of the bloom for constituting the necessary biomass of the algae which, in turn, is required for N_2 fixation. Along with the progress of the bloom, nutrient enrichment of the environment increases (the main source are the decaying of the algal cells and extra cellular metabolites) and is attributable to accessible exudates and recycling products (Devassy et al. 1978; Santhanam et al. 1994a, b). In the regions of the bloom, zooplankton biomass (as displacement volume) in the mixed layer was comparatively low. Copepods showed maximum numerical abundance and percentage contribution among zooplanktonic organisms followed by Chaetognaths. Available literature indicates that *Trichodesmium* is not a feed for most of the zooplankton, and this can be possibly the reason for the low mesozooplankton biomass in the mixed layer. The occurrence of *Trichodesmium* blooms in the open waters of Bay of Bengal demands more detailed studies targeting its ecological significance.

2.4.8 Hong Kong

Hong Kong is one of the worst HAB-affected areas in the world, with a high diversity of harmful and toxic algal species, especially dinoflagellates. Hong Kong is located in the north-eastern part of the subtropical South China Sea. It is surrounded by sea to the south, east and west. Junk Bay is located on the south of the Kowloon Peninsula, facing Hong Kong Island. In Hong Kong, a record algal bloom, caused by *Gymnodinium mikimotoi* and *Gyrodinium* sp. HK'98 (subsequently described as *Karenia digitata*) occurred in March and April 1998 (Lu and Goebel 2001). Majority of the fishes died in the affected cages, and there was severe economic loss caused by the HAB. Most of the known toxic or harmful algal species are dinoflagellates. Some common dinoflagellate species such as *Ceratium furca*, *Gonyaulax polygramma*, *Noctiluca scintillans*, *Heterocap satriquetra*, *Prorocentrum minimum*, *Prorocentrum sigmoides* and *Prorocentrum triestinum* frequently bloom in Hong Kong waters. Others, such as *Alexandrium catenella*, *Alexandrium tamarense*, *Gymnodinium mikimotoi*, *Gymnodinium* cf. *breve*, *Gymnodinium catenatum*, *Dinophysis caudata*, *Dinophysis acuminata* and *Gambierdiscus toxicus*, bloom only occasionally, but their toxic effects or potentially toxic and harmful effects are very significant. In 1988, a continuous bloom of *Gonyaulax polygramma* lasted for three and a half months from early February to May in Tolo Harbour. Red tides and algal blooms refer to the water discolouration caused by such high algal concentrations. Massive fish kills have been caused by oxygen depletion, and the occurrence of red tides at bathing beaches has inconvenienced the general public. Such harmful algal blooms (HABs) have become a serious problem in subtropical Hong Kong waters in the past two decades.

2.4.9 China

The frequency and spatial extension of HABs occurrence have been increasing steadily in the Chinese coastal waters since the 1980s, resulting considerable negative impact in marine aquaculture and causing enormous economic losses. The first documented HAB event in China was caused by *Noctiluca scintillans* (Macartney) (Kofoid and Sweazy 1921) and *Skeletonema costatum* (Greville) (Cleve 1873) in Zhejiang coastal waters in 1933. It killed marine organisms such as razor clams and other shellfish species (Fei 1952). From 1970s, HAB frequency increased three times every 10 years and sometimes led to large economic losses in Chinese coastal waters.

Wang et al. (2008) recorded a detailed account of HAB events during the period from 1980 to 2003 in South China Sea (SCS). They accounted those HABs-affected areas expanded and the frequency of HABs varied during this period, leading severe economic losses in aquaculture. The seasonal and annual variations, as well as causative algal species of HABs, are different among the four regions: the Pearl River

Estuary (China), the Manila Bay (the Philippines), the Masinloc Bay (the Philippines) and the western coast of Sabah (Malaysia). HABs occurred frequently during March–May in the northern region of SCS, May–July in the eastern region, July in the western region and year-round in the southern region. Among the species that cause HABs, *Noctiluca scintillans* (Macartney) (Kofoid and Sweazy 1921) dominated in the northern region and *Pyrodinium bahamense* (Plate, 1906) in the southern and eastern regions. Causative species also varied in different years for the entire SCS as follows: both *P. bahamense* (Plate, 1906) and *N. scintillans* (Macartney) (Kofoid and Sweazy 1921) were the dominant species during 1980–2003. Some species not previously recorded formed blooms during 1991–2003, including *Phaeocystis globosa* (*globosa* Scherffel 1899), *Scrippsiella trochoidea* (Stein) (Loeblich III 1976), *Heterosigma akashiwo*, (Y.Hada) (Y.Hada ex Y.Hara & M.Chihara 1987) and *Mesodinium rubrum* (Leegaard 1915). The variations in HABs were linked to a set of regional conditions, such as a reversed monsoon wind in the entire SCS, river discharges in the northern area, upwelling in Vietnam coastal waters during southwest winds and near Malaysia coastal waters during northeast winds and eutrophication from coastal aquaculture in the Pearl River Estuary.

Tang et al. (2006) had investigated the spatial and seasonal characteristics of HAB events in the southern Yellow Sea and East China Sea along Chinese coast for a prolonged period of time (1933–2004); covering South Yellow Sea and East China Sea, both are marginal seas of Pacific Ocean. Among the 435 HAB recorded events, the most frequent HAB occurrence area (FHA) is of the Yangtze River mouth, and another two FHA areas are located south of the Yangtze River estuary in the East China Sea. The time of HAB occurrence shifted from autumn (August–October) prior the 1980s, July–August in the 1980s, during May–July in the 1990s and May–June for the period of 2000–2004. HAB-causative species were as follows: *Noctiluca scintillans* (Macartney) (Kofoid and Sweazy 1921) and *Skeletonema costatum* (Greville) (Cleve 1873) were dominant causative species prior to 2000; and *Prorocentrum donghaiense* (Lu and Goebel 2001) was dominant from 2000 to 2004 and also caused large blooms in May. *Trichodesmium* sp. caused many HABs in autumn (August–October) prior to the 1980s with only one HAB between 1980 and 2004. The changes of the dominant HAB species may have affected the timings of HAB occurrence, as well as the increasing HAB-affected areas in recent years.

East China Sea (ECS) might be considered as the most vulnerable region for the occurrence of HABs in China, formed by both diatoms and dinoflagellates in diverse ways. This was mainly favoured by Changjiang diluted water (CDW) and currents from the open ocean (i.e. Taiwan warm current (TWC)), and these two have different roles in affecting the two types of algal blooms. Recently, Zhou et al. (2017) explained a better understanding of the mechanisms of HABs in the East China Sea. They pointed out precisely that phosphate and silicate are the major factors which directly initiated the diatom bloom, while dissolved inorganic nitrogen (DIN), temperature and turbidity are the factors that influence the dinoflagellate bloom. CDW has a high concentration of nitrate as well as silicate is essential for the diatom bloom, while the intrusion of the TWC (mainly Kuroshio subsurface water that is rich in phosphate at the bottom) is critical for the occurrence of the dinoflagellate bloom.

The sea area adjacent to the Changjiang River estuary is the most notable region for harmful algal blooms (HABs) of dinoflagellates in China. Large-scale blooms of dinoflagellates (i.e. *Prorocentrum donghaiense* (Lu and Goebel 2001) and *Karenia mikimotoi* (Miyake and Kominami ex Oda) (Gert Hansen & Ø.Moestrup in N. Daugbjerg, G. Hansen, J. Larsen, & Ø. Moestrup 2000)) started to appear in this region at the beginning of the twenty-first century, and the mechanisms for the occurrence of dinoflagellate blooms have become an intriguing issue since then. In previous studies, an apparent succession of microalgal blooms from diatoms to dinoflagellates, due to their competition and the different adaptive strategies to environmental variations, has been found in late spring. Therefore, a better understanding on the succession of different microalgal blooms would help to elucidate the mechanisms of the large-scale dinoflagellate blooms.

To get insight into the seasonal succession mechanisms of microalgal blooms, numerical model is an important tool to extract information from the complex field investigation data (Mutshinda et al. 2013). The simulation results could compensate the shortage of in situ data by offering detailed temporal changes and spatial distributions of the algal blooms. Recently, a zero-dimensional numerical model was established by Zhou et al. (2017) to simulate the succession of microalgal blooms based on the field investigation data during 2005 and 2011. The model could well reproduce dynamics of the diatom and dinoflagellates blooms. Using this model, analyses were performed under different scenarios to analyse the effects of temperature, light intensity and nutrients on the succession of microalgal blooms. The results suggest that temperature and light have little effects on the succession. Phosphorous stress is the most critical factor controlling the succession of microalgal blooms, while nitrate plays an important role in affecting the scale of dinoflagellate bloom. The results further improved the current understandings on the mechanisms of large-scale dinoflagellate blooms in the East China Sea and would help to develop prevention strategies against HABs in this region.

In this study, a zero-dimensional numerical model was established to simulate the dynamics and succession of diatom and dinoflagellate blooms in the coastal waters adjacent to the Changjiang River estuary, where the large-scale blooms of dinoflagellates became a notable phenomenon from the beginning of the twenty-first century. Parameters of the model were optimized based on the field observation data in 2005, and the model could well reproduce the dynamics and succession of microalgal blooms in the year 2005 and 2011. Using the model, the roles of different environmental factors, i.e. temperature, light intensity and nutrients, were examined under different scenarios. It can be implied that phosphate is the most critical factor for the decline of diatom bloom and the succession from diatom blooms to dinoflagellate blooms. Nitrate is important in determining the intensity of dinoflagellate blooms. So far silicate is still rich in seawater and not an important factor affecting the intensity of diatom blooms. Both temperature and light are not critical factors for the succession of microalgal blooms. But then crease of temperature would benefit the growth of dinoflagellates and promote the intensity of dinoflagellate blooms.

2.4.10 Japan

Harima-Nada, located in the eastern part of the Seto Inland Sea, is a semi-enclosed sea (size of 3400 km^2; average depth of 26 m). Due to industrialization and urbanization, this area becomes one of the most seriously impacted areas by eutrophication in Japanese coastal areas (Yamamoto 2003). It is a major fishing ground including aquaculture of fish, bivalves and seaweeds and is also one of the most industrially developed areas in Japan.

Nishikawa et al. (2011) had monitored for a long period of time (over 35 years) to establish the long-term variations of the harmful diatom *Eucampia zodiacus*, in Harima-Nada, a causative organism for bleaching of aquacultured "nori" (*Porphyra yezoensis*). *E. zodiacus* cells were detected every year, and seasonal cell densities tended to be higher from January to April. The proportion of *E. zodiacus* to the total phytoplankton cell density has increased in recent years, because the abundances of *E. zodiacus* started to increase in the mid-1990s. However, cell densities of the total phytoplankton in the 1980s and thereafter appeared to be lower in comparison to the 1970s. During the 35-year period, there were two significant long-term changes, i.e. an increase in winter water temperatures and a decrease in the concentration of dissolved inorganic nitrogen. The present results suggest that the shift of environmental conditions is more advantageous to the growth of *E. zodiacus*, which contributed to the domination of *E. zodiacus* abundance in recent years.

Nishikawa et al. (2014) intensively studied the long-term changes (during 1973–2008) in three species of *Chattonella* spp. (*Chattonella antiqua* (Y. Hada), *Chattonella marina* (Subrahmanyan) and *Chattonella ovata*) in the context of environmental factors in Harima-Nada, eastern Seto Inland Sea, Japan. The harmful alga *Chattonella* is a genus of marine raphidophytes associated with red tides, and the three species, *Chattonella antiqua*, *C. marina* and *C. ovata* (Raphidophyceae), have been classified based on their morphological characteristics and genetic diversity.

Long-term trends in the dynamics of *Chattonella* populations were considered to relate to environmental factors such as nutrient concentrations and water temperature. During the 1970s to the early 1980s, enrichment of nutrient levels has contributed to the high cell density and large-scale red tides of *Chattonella* spp. in Harima-Nada. However, nutrient levels exhibited a decreasing trend thereafter, and it is thought that *Chattonella* spp. cannot form large-scale blooms under the present conditions. After the mid-1990s, the occurrence period of vegetative cells of *Chattonella* spp. has been several weeks or 1 month, and the appearance frequency of *Chattonella* spp. has increased in the northern coastal region. However, both the cell density and the spatial distribution have become lower and smaller than those in the previous decades. It is notable to refer that germination timing of *Chattonella* cysts has become earlier mainly due to the increase in water temperature and the chances of vegetative growth have also increased, especially at the northern coast where most of large rivers discharge into the Harima-Nada. In addition, the results revealed that presence of less number of diatoms was also important cause for the high numerical abundance of *Chattonella* spp. in this marine environment.

2.4.11 Northern Arabian Sea along Karachi, Pakistan

Sewage water from Karachi and agricultural runoff from adjacent lands put lots of pollutants and nutrients in coastal waters of Karachi. For monitoring study along the Karachi coast, phytoplankton samples were collected fortnightly from polluted as well as pristine localities by using a phytoplankton net of 40 μm mesh size. Physico-chemical factors such as temperature, salinity, pH and rainfall were also documented. Samples were analysed qualitatively using phase-contrast microscope. Unusual, non-toxic and irregular blooms of a pollution indicator microalga, *Synedra acus*, have been frequently observed, suppressing existence of all other phytoplankton.

2.4.11.1 Introduction

Karachi harbour receives large amount of industrial and sewage effluents of Karachi city through Lyari River (Saifullah et al. 2000, 2002a, b; Qureshi et al. 2001; Mashiatullah et al. 2009; Qadri et al. 2011; Nergis et al. 2012), which finally enters into the Arabian Sea, causing marine pollution in the vicinity of Karachi.

Composition of phytoplankton community structure in the region is largely influenced with the continuous inflow of several pollutants along with agricultural runoff. Seasonal southwest and northeast monsoonal regimes also have profound impact on the nutrient cycling in the marine environment. These have also distinct role in controlling the seasonal variations in phytoplankton communities. Occurrence of algal blooms in new geographical areas with their increasing frequency has been recorded which is a matter of global concern. These blooms are being mainly concerned to increased pollution as well as impact of climatic changes. Majority of algal blooms reported in Indian Ocean has been innocuous in nature, excepting few cases which are responsible for fish kill due to formation of anoxic conditions (Rabbani et al. 1990). *Prorocentrum minimum* was the causative agent in Gwadar Bay (Rabbani et al. 1990) for mass mortality of fishes, and this was an exclusive report from Pakistan coast.

Appearance of algal blooms can be suspected along Karachi coast under such seasonal changes of nutrient and pollutants input. But the appearances of any bloom along Karachi coast have not been reported in the past. Algal blooms in coastal waters of Karachi might have gone unnoticed, because of producing no distinctive colour or due to lack of careful monitoring and sampling during such blooms. The case study was carried out by Luqman et al. (2015) to identify diatom blooms coastal waters of North Arabian Sea along Karachi, Pakistan, covering different seasons. Study area was considered in the coastal waters in the vicinity of Karachi including Karachi Harbour (Lat.24°.48′N, Long.66°.58′E). Three study sites were selected considering both polluted and unpolluted localities, namely, station 1 (outside Manora channel, cleaner locality outside harbour (24.77203°N 66.99386°E)), station 2 (near Manora channel, within polluted Karachi Harbour (24.79333°N 66.98248°E)) and station 3 (backwaters of sandspit, within polluted Karachi Harbour (24.81694°N 66.96746°E)). The presence of blooms was correlated with seasonal changes, monsoonal conditions, salinity, temperature (°C) and monthly and annual rainfall. Relationship between these bloom and salinity, temperature and rainfall has been shown in graphical forms (Fig. 2.11).

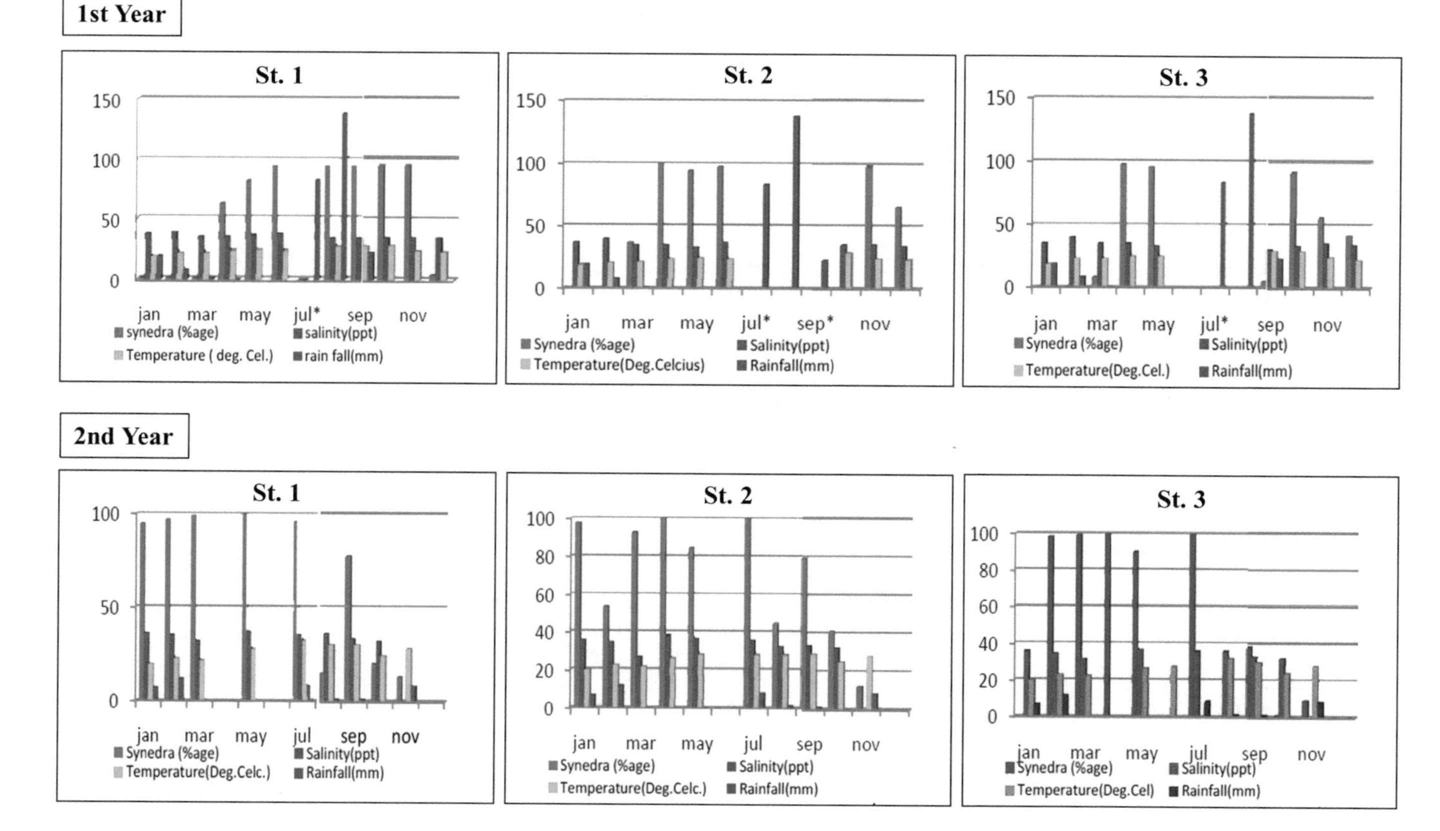

Fig. 2.11 Occurrence of *Synedra acus* at three different stations in 2 consecutive years

2.4.11.2 Important Findings

During both years of monitoring, the diatom *Synedra acus* was present exceptionally in high percentage at all three locations. This is long, needle-like, epiphytic, motile, benthic and pollution indicator micro algae that exist solitary or in radiate colonies. The bloom of this species, which naturally remains restricted to particular polluted location associated with certain organic pollutants (Jafari and Gunale 2005), is the first ever reported bloom of any pollution indicator microalgae from coastal waters of Pakistan, bordering northern Arabian Sea.

During first year of monitoring at station 1, which represents relatively coastal waters of Karachi outside Manora channel, this species showed its bloom from April to November approaching up to 90% frequency of occurrence from June to November. *Synedra acus* was not detected in the samples collected during January and was almost negligible (less than 1%) during February and March and less than 5% during December. The absence or nominal presence of this species was observed during January to March and December, 1990, which is calm period of northeast monsoon season with minimal wave action, less turbidity and turbulence. At station 2, *Synedra acus* bloom was observed during March, April, May, June, November and December. During July, August and September, samples were not collected. *S. acus* was not found during October at this sampling site, but during this month at station 1, it has shown more than 90% frequency of occurrence. This showed that the bloom observed at station 1 was limited to coastal waters and did not extend inside the harbour. This observation indicates that our open waters are not as cleaner and pollution free as were thought. Now they are harbouring pollution indicator algae. At station 3, *S. acus* bloom was found during April, May, October, November and December, whereas no bloom was recorded in September. Similar irregular pattern was observed during 2nd year of monitoring at all three stations. However the contrary observation was the appearance of *S. acus* bloom in January and February. October and November of the 2nd year has shown low percentage of occurrence of *S. acus*, whereas in the preceding year, these months had shown bloom of this species. During 1st year of monitoring, *S. acus* did not bloom in northeast monsoon season, but in the coming year it has shown blooms even during northeast monsoon season contradicting all the assumptions made on the basis of prior results. These blur data of *S. acus* bloom showed no clear relation with any of the oceanography parameters under consideration, i.e. salinity, temperature and rainfall. This appearance may be caused by presence of persistent organic pollutants like organochlorine pesticides (DDT, HCH, etc.) in these coastal waters. The relationship of *S. acus* with polychlorinated biphenyl (PCB) along with some other pinnate diatom species has been reported in the United States (Fitzgerald and Steuer 2005). High level of PCBs has also been identified in waters (Munshi et al. 2005), indicating towards Karachi coastal pollution factor which might have helped this pollution indicator species not even to exist but to bloom and out competing all other microalgal cells.

2.4.11.3 Conclusion

Non-toxic blooms of *S. acus* sporadically appeared in coastal waters of Karachi. This irregular appearance of pollution indicator diatom was not related with any environmental parameters such as salinity, temperature and rainfall. These results reflect the presence of predictable factors (nutrients, environmental pollutants as well as climatic conditions) that might have governed *S. acus* to bloom in this part of the northern Arabian Sea. Presence of this pollution specific microalga suggests strict and immediate measures to curb the pollutants input into marine waters from the industrial and metropolitan city of Karachi.

2.4.12 EEZ of India

An extensive account of the occurrence, frequency, intensity and spatial coverage of harmful algal blooms in the EEZ of India were documented by Padmakumar et al. (2012). Out of 80 occurrences of algal blooms during 1998–2010, dinoflagellates were dominant (31), followed by 27 by cyanobacteria (27) and diatoms (18) along with very few cases by raphidophyte and haptophyte. Potentially toxic microalgae recorded from the Indian waters were *Alexandrium* spp., *Gymnodinium* spp., *Dinophysis* spp., *Coolia monotis, Prorocentrum lima*, and *Pseudo-nitzschia* spp., were the bloom forming agents from the Indian waters. Among the 422 species of microalgae from EEZ of India, diatoms were dominated by 219 species, followed by 179 species of dinoflagellates, 16 species of blue-green algae and 8 other groups (silicoflagellates, chlorophytes, coccolithophorids, raphidophytes, haptophytes and prasinophytes). The bloom occurring dinoflagellate *Noctiluca scintillans* was the dominant and frequently occurring species in the South Eastern Arabian Sea (SEAS) during the summer monsoon and green *Noctiluca scintillans* in the North Eastern Arabian Sea (NEAS) during the winter cooling. The other bloom-forming dinoflagellates are *Cochlodinium* sp., *Gymnodinium* sp., *Gonyaulax* sp. and *Ceratium* spp. The authors ascertained that the *Trichodesmium* bloom occurred annually, but the *Noctiluca* bloom appeared at irregular intervals.

2.5 Conclusion

This chapter provides an overview of important drivers for the occurrence of algal bloom and their adverse impact on biotic and abiotic components. This is evident that there exist sharp differences in regional scale occurrences and the causative species of algal blooms, frequency, intensity, spatial coverage and their adverse impact for different tropical coastal regions. Bloom causative agents are naturally driven due to physical forcing such as monsoonal influence, riverine discharge, seasonal upwelling and wind direction and speed. Eddies and cyclones can also be considered as a triggering factor for provoking bloom. Besides these factors, variations environmental factors (temperature, salinity, irradiance, water stability, nutrient-enriched waters) as well as biotic factor (such as prey availability) are also important that influence bloom formation.

A review of bloom occurrences in coastal regions of India indicates their predominance along the west coast of India especially the southern part. Majority of the blooms reported along the west coast of India are caused by dinoflagellates, whereas diatom blooms prevail along the east coast. There have been 39 causative species responsible for blooms, of which *Noctiluca scintillans* and *Trichodesmium erythraeum* are the most common. Reporting of massive fish mortality in Indian waters has been associated with the blooming of *Cochlodinium polykrikoides*, *Karenia brevis*, *Karenia mikimotoi*, *Noctiluca scintillans*, *Trichodesmium erythraeum*, *Trichodesmium thiebautii* and *Chattonella marina*. Majority of the blooms occurred during withdrawal of the southwest monsoon and pre-monsoon period. In Indian waters, this process is mainly influenced by seasonal upwelling and monsoonal forcing that causes high riverine discharge resulting in nutrient-enriched waters that provides a competitive edge for blooming of phytoplankton species. Since these algal bloom outbreaks are sporadic and unpredictable, regular monitoring of bloom-prone areas will provide significant insights into bloom dynamics and thus provide a useful tool for bloom monitoring as a step ahead.

References

Adhavan, D., Kamboj, R. D., Chavdaand, D. V., & Bhalodi, M. M. (2014a). Status of intertidal biodiversity of Narara Reef Marine National Park, Gulf of Kachchh, Gujarat. *Journal of Marine Biology and Oceanography, 3*(3), 2.

Adhavan, D., Kamboj, R. D., Marimuthu, N., et al. (2014b). Seasonal variation and climate change influence coral bleaching in Pirotan Island, Gulf of Kachchh Marine National Park, Gujarat. *Currrent Science, 107*(11), 1780–1781.

Adhikary, S. P., & Sahu, J. (1992). Studies on the Trichodesmium bloom of Chilka Lake, East Coast of India. *Phykos, 30*, 101–107.

Ahmed, S., Arakawa, O., & Onoue, Y. (1995). Toxicity of cultured *Chattonella marina*. In P. Lassus, G. Arzul, E. Erad, P. Geniten, & C. Marciallou (Eds.), *Harmful algal blooms* (pp. 499–504). Paris: Techinique at documentation-Lavoiser Intercept Ltd.

Alagaraja, K., Kurup, K. N., Srinath, M., & Balakrishnan, G. (1992). *Analysis of marine landings in India- a new approach* (CMFRI Special Publication, Vol. 10, p. 42). Cochin: Central Marine Fisheries Research Institute.

Al-Azri, A., Al-Hashmi, K., Goes, J., Gomes, H., Rushdi, A. I., Al-Habsi, H., et al. (2007). Seasonality of the bloom-forming heterotrophic dinoflagellate *Noctiluca scintillans* in the Gulf of Oman in relation to environmental conditions. *International Journal of Oceans and Oceanography, 2*(1), 51–60.

Allen, J. I., Anderson, D., Burford, M., Dyhrman, S., Flynn, K., Glibert, P. M., Granéli, E., Heil, C., Sellner, K., Smayda, T., & Zhou, M. (2006). *Global ecology and oceanography of harmful algal blooms, harmful algal blooms in eutrophic systems* (P. Glibert, Ed., GEOHAB report 4, p. 74). Paris/Baltimore: IOC and SCOR.

Anas, A., Sheeba, V. A., Jasmin, C., Gireeshkumar, T. R., Mathew, D., Krishna, K., Nair, S., Muraleedharan, K. R., & Jayalakshmy, K. V. (2018). Upwelling induced changes in the abundance and community structure of archaea and bacteria in a recurring mud bank along the southwest coast of India. *Regional Studies in Marine Science, 18*, 113–121.

Andreae, M. O., & Crutzen, P. J. (1997). Atmospheric aerosols: Biogeochemical sources and role in atmospheric chemistry. *Science, 276*(5315), 1052–1058.

Archer, S. D., Widdicombe, C. E., Tarran, G. A., Rees, A. P., & Burkill, P. H. (2001). Production and turnover of particulate dimethylsulphoniopropionate during a coccolithophore bloom in the northern North Sea. *Aquatic Microbial Ecology, 24*(3), 225–241.

Arhonditsis, G., Tsirtsis, G., & Karydis, M. (2002). The effects of episodic rainfall events to the dynamics of coastal marine ecosystems: Applications to a semi-enclosed gulf in the Mediterranean Sea. *Journal of Marine Systems, 35*, 183–205.

Arun Kumar, M., Karthik, R., Sai Elangovan, S., & Padmavati, G. (2012). Occurrence of *Trichodesmium erythraeum* bloom in the coastal waters of south Andaman. *International Journal of Current Research, 11*, 281–284.

Arun Kumar, M., Padmavati, G., & Pradeep, H. D. (2015). Occurrence of *Trichodesmium erythraeum* (Cyanophyte) bloom and its effects on the fish catch during April 2013, in the Andaman Sea. *Applied Environmental Research, 37*, 49–57.

Barría de Cao, M. S., Beight, M., & Piccolo, C. (2005). Temporal variability of diversity and biomass of tintinnids (Ciliophora) in Southeastern Atlantic temperate estuary. *Journal of Plankton Research, 27*(11), 1103–1111.

Begum, M., Sahu, B. K., Das, A. K., Vinithkumar, N. V., & Kirubagaran, R. (2015). Extensive *Chaetocero scurvisetus* bloom in relation to water quality in Port Blair Bay, Andaman Islands. *Environmental Monitoring and Assessment, 187*, 1–14.

Bhat, S. R., & Verlencar, X. N. (2006). Some enigmatic aspects of marine cyanobacterial genus, *Trichodesmium*. *Current Science, 91*, 18–19.

Bhimachar, B. S., & George, P. C. (1950). Abrupt setback in the fisheries of the Malabar and Kanara coasts and red water phenomenon and their probable cause. *Proceedings of the Indian Academy of Sciences, 31*, 339–350.

Biswas, S. N., Godhantaraman, N., Sarangi, R. K., Bhattacharya, B. D., Sarkar, S. K., & Satpathy, K. K. (2013). Bloom of *Hemidiscus hardmannianus* (Bacillariophyceae) and its impact on water quality and plankton community structure in a mangrove wetland. *Clean – Soil, Air, Water, 41*(4), 333–339.

Biswas, S. N., Rakshit, D., Sarkar, S. K., Sarangi, R. K., & Satpathy, K. K. (2014). Impact of multispecies diatom bloom on plankton community structure in Sundarban mangrove wetland, India. *Marine Pollution Bulletin, 85*, 306–311.

Blackburn, S. I., McCausland, M. A., Bolch, C. J. S., Newman, S. J., & Jones, G. J. (1996). Effect of salinity on growth and toxin production in cultures of the bloom-forming cyanobacterium Nodularia spumigera from Australian waters. *Phycologia, 36*(6), 511–522.

Blanchard, J. L., Jennings, S., Holmes, R., Harle, J., Merino, G., Allen, J. I., et al. (2012). Potential consequences of climate change for primary production and fish production in large marine ecosystems. *Philosophical Transactions of the Royal Society of London. Series B, Biological Sciences, 367*, 2979–2989.

Bopp, L., Aumont, O., Belviso, S., & Monfray, P. (2003). Potential impact of climate change on marine dimethyl sulfide emissions. *Tellus, 55B*, 11–22.

Cameron-Smith, P., Elliott, S., Maltrud, M., Erickson, D., & Wingenter, O. (2011). Changes in dimethyl sulfide oceanic distribution due to climate change. *Geophysical Research Letters, 38*, L07704.

Capone, D. G., Zehr, J. P., Paerl, H. W., Bergman, B., & Carpenter, E. J. (1997). *Trichodesmium*, a globally significant marine bacteria. *Science, 276*, 1221–1229.

Carmichael, W. W. (1992). Cyanobacterial secondary metabolites – The cyanotoxins. *The Journal of Applied Bacteriology, 724*, 45–459.

Carslaw, K. S., Boucher, O., Spracklen, D. V., Mann, G. W., Rae, J. G. L., Woodward, S., & Kulmala, M. (2010). A review of natural aerosol interactions and feedbacks within the Earth system. *Atmospheric Chemistry and Physics, 10*(4), 1701–1737.

Chacko, P. I. (1942). An unusual incidence of mortality of marine fauna. *Current Science, 11*, 404.

Chang, J., Chiang, K. P., & Gong, G. C. (2000). Seasonal variation and cross-shelf distribution of the nitrogen-fixing cyanobacterium, *Trichodesmium*, in the southern East China Sea. *Continental Shelf Research, 20*, 479–492.

Charlson, R. J., Lovelock, J. E., Andreaei, M. O., & Warren, S. G. (1987). Oceanic phytoplankton, atmospheric sulphur, cloud. *Nature, 326*(6114), 655–661.

Chellappa, S. I., Marinho, I. R., & Chellappa, N. T. (2004). Freshwater phytoplankton assemblages and the bloom of toxic *Cyanophyceae* of Campo Grande reservoir of Rio Grande do Norte State of Brazil. *Indian Hydrobiology, 7*, 151–171.

Chidambaram, K., & Menon, K. (1945). Correlation of the west coast (Malabar and South Kanara) fisheries with plankton and certain oceanographical features. *Proceedings of the Indian Academy of Sciences, 31*, 252–286.

Chorus, I., & Bartram, J. (1999). *Toxic cyanobacteria in monitoring and management. E and FN Spon* (416 pp). London: An Imprint of Routledge.

Clarke, K. R., & Warwick, R. M. (1998). Quantifying structural redundancy in ecological communities. *Oecologia, 113*(2), 278–289.

Clarke, K. R., & Warwick, R. M. (2001). *Change in marine communities: An approach to statistical analysis and interpretation* (2nd ed.p. 171). Plymouth: PRIMER-E.

D'Silva, M. S., Anil, A. C., Naik, R. K., & D'Costa, P. M. (2012). Algal blooms: A perspective from the coasts of India. *Natural Hazards, 63*, 1225–1253.

Davidson, K., Miller, P., Wilding, T. A., Shutler, J., Bresnan, E., Kennington, K., & Swan, S. (2009). A large and prolonged bloom of *Karenia mikimotoi* in Scottish waters in 2006. *Harmful Algae, 8*, 349–361.

Devassy, V. P. (1987). *Trichodesmium* red tides in the Arabian Sea. In T. S. S. Rao (Ed.), *Contributions in marine sciences: A special volume to felicitate Dr. S. Z. Qasim Sastyabdapurtl on his sixtieth birthday* (pp. 61–66). Dona Paula: National Institute of Oceanography.

Devassy, V. P., & Nair, S. R. S. (1987). Discolouration of waters and its effect on fisheries along the Goa coast. *Mahasagar, 20*, 121.

Devassy, V. P., Bhatrarhiri, P. M. A., & Qasim, S. Z. (1978). *Trichodesmium* phenomenon. *Indian Journal of Marine Science, 73*, 168–186.

Dharani, G., Abdul Nazar, A., Kanagu, L., Venkateshwaran, P., Kumar, T., Ratnam, K., Venkatesan, R., & Ravindran, M. (2004). On the recurrence of *Noctiluca scintillans* bloom in Minnie Bay, Port Blair: Impact on water quality and bioactivity of extracts. *Current Science, 87*, 990–994.

Dippner, J. W., Nguyen-Ngoc, L., Doan-Nhu, H., & Subramaniam, A. (2011). A model for the prediction of harmful algae blooms in the Vietnamese upwelling area. *Harmful Algae, 10*(6), 606–611.

Doney, S. C., Fabry, V. J., Feely, R. A., & Kleypas, J. A. (2009). Ocean acidification: The other CO_2 problem. *Annual Review of Marine Science, 1*, 169–192.

Duguay, L. E., Monteleone, D. M., & Monteleone, C. E. (1989). Abundance and distribution of zooplankton and Ichthyoplankton in Great South Bay, New York: During the brown tide outbreaks of 1985 and 1986. In E. M. Cosper, V. M. Bricelj, & E. J. Carpenter (Eds.), *Novel phytoplankton blooms* (pp. 600–623). Berlin: Springer.

Dwivedi, R. M., Chauhan, R., Solanki, H. U., Raman, M., Matondkar, S. G. P., Madhu, V., & Meenakumari, B. (2012). Study of ecological consequence of the bloom (*Noctiluca miliaris*) in off shore waters of the Northern Arabian Sea. *Indian Journal of Geo Marine Sciences, 41*(4), 304–313.

Eashwar, M., Nallathambi, T., Kuberaraj, K., & Govindarajan, G. (2001). *Noctiluca* blooms in Port Blair Bay, Andamans. *Arya, 1105*, 1–10.

Elangovan, S. S., Arun Kumar, M., Karthik, R., Siva Sankar, R., Jayabarathi, R., & Padmavati, G. (2012). Abundance, species composition of microzooplankton from the coastal waters of Port Blair, South Andaman Island. *Aquatic Biosystems, 8*, 20.

Elbrächter, M. & Qi, Y. Z. (1998). Aspects of Noctiluca (Dinophyceae) population dynamics. In D. M. Anderson et al.(Eds.), *Physiological ecology of harmful algal blooms* (NATO ASI Series, Vol. G41, pp. 315–335). Berlin: Springer.

Endo, M., Onoue, Y., & Kuroki, A. (1992). Neurotoxin induced disorder and its role in the death of fish exposed to *Chattonella marina*. *Marine Biology, 112*, 371–376.

Falconer, I. R., Burch, M. D., Steffensen, D. A., Choice, M., & Coverdale, O. R. (1994). Toxicity of the blue-green alga (cyanobacterium) *Microcystis aeruginosa* in drinking water to growing pigs, as an animal model for human injury and risk assessment. *Environmental Toxicology, 9*(2), 131–139.

Fay, P. (1983). *The blue greens* (Studies in biology N°160). London: Edward Arnold.

Federico, A., Sarma, S. S. S., & Nandini, S. (2007). Effect of mixed diets (cyanobacteria and green algae) on the population growth of the cladocerans *Ceriodaphnia dubia* and *Moina macrocopa*. *Aquatic Ecology, 41*, 579–585.

Fei, H. (1952). The cause of red tides. *Science and Art 22*, 1–3 (in Chinese).

Fernandes, L. F., Zehnder-Alves, L., & Bassfeld, J. C. (2001). The recently established diatom *Coscinodiscus wailesii* (Coscinodiscales, Bacillariophyta) in Brazilian waters. I: Remarks on morphology and distribution. *Phycological Research, 49*, 89–96.

Fistarol, G. O., Legrand, C., & Granéli, E. (2005). Allelopathic effect on a nutrient-limited phytoplankton species. *Aquatic Microbial Ecology, 41*(2), 153–161.

Fitzgerald, S. A., & Steuer, J. J. (2005). Association of PCBs with live algae and total lipids in rivers. *The Science of the Total Environment, 354*, 60–74.

Gabric, A. J., Simó, R., Cropp, R. A., Hirst, A. C., & Dachs, J. (2004). Modeling estimates of the global emission of dimethylsulfide under enhanced greenhouse conditions. *Global Biogeochemical Cycles, 18*, GB2014.

Gayoso, A. M. (1999). Seasonal succession patterns of phytoplankton in the Bahía Blanca Estuary (Argentina). *Botanica Marina, 42*, 367–375.

Ghadouani, A., Pinel-Alloul, B. B., & Prepas, E. E. (2006). Could increase cyanobacterial biomass following forest harvesting cause a reduction in zooplankton body size structure? *Canadian Journal of Fisheries and Aquatic Sciences, 63*, 2308–2317.

Glibert, P. M., Anderson, D. M., Gentien, P., Granéli, E., & Sellner, K. G. (2005). The global, complex phenomena of harmful algal blooms. *Oceanography, 18*(2), 130–141.

Gopal, B., & Chauhan, M. (2006). Biodiversity and its conservation in the Sundarban mangrove ecosystem. *Aquatic Sciences, 69*, 338–354.

Gosselin, S., Fortier, L., & Gagne, J. A. (1989). Vulnerability of marine fish larvae to the toxic dinoflagellate *Protogonyaulax tamerensis*. *Marine Ecology Progress Series, 57*, 1–10.

Granéli, E., & Hansen, P. J. (2006). Allelopathy in harmful algae: A mechanism to compete for resources? In E. Granéli & J. T. Turner (Eds.), *Ecology of harmful algae* (pp. 189–201). Berlin: Springer.

Granéli, E., Weberg, M., & Salomon, P. S. (2008). Harmful algal blooms of allelopathic microalgal species: The role of eutrophication. *Harmful Algae, 8*, 94–102.

Gypens, N., & Borges, A. V. (2014). Increase in dimethylsulfide (DMS) emissions due to eutrophication of coastal waters offsets their reduction due to ocean acidification. *Frontiers in Marine Science, 1*, 4.

Halegraeff, D. M., Anderson, A., Cembella, D., & Envlodsen, H. O.. (1995). *Manual on harmful marine microalgae* (IOC manuals and guides, Vol. 33, pp. 550). Rome: UNESCO.

Hall, S. J., & Greenstreet, S. P. (1998). Taxonomic distinctness and diversity measures: Responses in fish communities. *Marine Ecology Progress Series, 166*, 227–229.

Hallegraeff, G., & Gollasch, S. (2006). Anthropogenic introductions of microalgae. In *Ecology of harmful algae* (pp. 379–390). Berlin/Heidelberg: Springer.

Hobson, P., Burch, M., & Fallowfield, H. J. (1999). Effect of total dissolved solids and irradiance on growth and toxin production by *Nodularia spumigera*. *Journal of Applied Phycology, 11*, 551–558.

Hornell, J. (1917). A new protozoan cause of widespread mortality among marine fishes. *Madras Fisheries Investment Bulletin, 1*, 53–56.

Hubbart, B., Pitcher, G. C., Krock, B., & Cembella, A. D. (2012). Toxigenic phytoplankton and concomitant toxicity in the mussel Choromytilus meridionalis off the west coast of South Africa. *Harmful Algae, 20*, 30–41.

Ishimatsu, A., Maruta, H., Tsuchiyama, T., & Ozaki, M. (1990). Respiratory, ionoregulatory and cardiovascular responses of the yellow tail *Seriola quinqeradiata* on exposure to the red tide plankton *Chattonella*. *Nippon Suisan Gakkaishi, 56*, 189–199.

Jacob, P. K., & Menon, M. D. (1948). Incidence of fish mortality on the west coast. *Journal of the Bombay Natural History Society, 47*, 455.

Jafari, N. G., & Gunale, V. R. (2005). Hydrobiological study of algae of an urban freshwater river. *Journal of Applied Science and Environment Management, 10*, 153–158.

John, D. M., Whitton, B. A., & Brook, A. J. (2002). *The freshwater algal Flora of the British Isles. An Identification guide to freshwater and terrestrial algae* (p. 702). Cambridge: Cambridge University Press/Natural History Museum.

Jugnu, R., & Kripa, V. (2009). Effect of *Chattonella marina* [(Subrahmanyan) Hara etChihara 1982] bloom on the coastal fishery resources along Kerala coast, India. *Indian Journal of Geo-marine Sciences, 38*(1), 77–78.

Jyothibabu, R., Madhu, N. V., Murukesh, N., Haridas, P. C., Nair, K. K. C., & Venugopal, P. (2003). Intense blooms of *Trichodesmium erythraeum* (Cyanophyta) in the open waters along east coast of India. *Indian Journal of Marine Sciences, 32*(2), 165–167.

Karthik, R., & Padmavati, G. (2017). Temperature and salinity are the probable causative agent for the *Trichodesmium erythraeum* (Cyanophyceae) algal bloom on the Burmanallah coastal waters of South Andaman Island. *World Applied Sciences Journal, 35*(8), 1271–1281.

Karthik, R., Arun Kumar, M., Sai Elangovan, S., Sivasankar, R., & Padmavati, G. (2012). Phytoplankton abundance and diversity in the coastal waters of Port Blair, South Andaman Island in relation to environmental variables. *Journal of Marine Biological Oceanography, 1*, 1–6.

Karthik, R., Arun Kumar, M., & Padmavati, G. (2014). Silicate as the probable causative agent for the periodic blooms in the coastal waters of south Andaman Sea. *Applied Environmental Research, 36*, 37–45.

Karunasagar, I., & Karunasagar, I. (1992). *Gymnodinium nagasakiense* red tide off Someshwar, West coast of India and mussel toxicity. *Journal of Shellfish Research, 11*, 477.

Keller, M. D. (1989). Dimethyl sulfide production and marine phytoplankton: The importance of species composition and cell size. *Biological Oceanography, 6*(5–6), 375–382.

Kim, J. H., Kim, J. H., Wang, P., Park, B. S., & Han, M. S. (2016). An improved quantitative real-time PCR assay for the enumeration of Heterosigmaakashiwo (Raphidophyceae) cysts using a DNA debris removal method and a cyst-based standard curve. *PloS one, 11*(1), e0145712.

Kloster, S., Six, K. D., Feichter, J., Maier-Reimer, E., Roeckner, E., Wetzel, P., et al. (2007). Response of dimethylsulfide (DMS) in the ocean and atmosphere to global warming. *Journal of Geophysical Research, 112*, G03005.

Kofoid, C. A., & Sweazy, M. (1921). The free living unarmoured Dionoflagellata. *Memoirs of the University of California, 5*, 1–562.

Kononen, K., & Leppänen, J. M. (1997). Patchiness, scales and controlling mechanisms of cyano-bacterial blooms in the Baltic Sea: application of a multi-scale research strategy. In M. Kahru & C. W. Brown (Eds.), *Monitoring algal blooms: New techniques for detecting large-scale environmental change* (pp. 63–84). Austin: Landes Bioscience.

Koya, K. P. S., & Kaladharan, P. (1997). *Trichodesmium* bloom and mortality of *Canthigaster margaritatus* in the Lakshadweep Sea. *Marine Fisheries Information Service Technical and Extension Series, 147*, 14.

Krishnan, A. A., Krishnakumar, P. K., & Rajagopalan, M. (2007). *Trichodesmium erythraeum* (EHR) bloom along the Southwest coast of India (Arabian Sea) and its impact on trace metal concentrations in seawater. *Estuarine, Coastal and Shelf Science, 71*, 641–646.

Kubanek, J., Hicks, M. K., Naar, J., & Villareal, T. A. (2005). Does the red tide dinoflagellate *Karenia brevis* use allelopathy to outcompete other phytoplankton? *Limnology and Oceanography, 50*, 883–895.

Lana, A., Bell, T. G., Simó, R., Vallina, S. M., Ballabrera-Poy, J., Kettle, A. J., et al. (2011). An updated climatology of surface dimethlysulfide concentrations and emission fluxes in the global ocean. *Global Biogeochemical Cycles, 25*, 1–17.

Leegaard, C. (1915). UntersuchungenübereinigePlanktonciliaten des Meeres. *Nytt Mag Naturvid, 53*, 1–37.

Legrand, C., Rengefors, K., Fistarol, G. O., & Granéli, E. (2003). Allelopathy in phytoplankton – Biochemical, ecological and evolutionary aspects. *Phycologia, 42*(4), 406–419.

Lewis, W. M. J. (1986). Evolutionary interpretations of allelochemical interactions in phytoplankton algae. *American Naturalist, 127*, 184–194.

Long, R. A., & Azam, F. (2001). Antagonistic interactions among marine pelagic bacteria. *Applied and Environmental Microbiology, 67*, 4975–4983.

Lu, D., & Goebel, J. (2001). Five red tide species in genus *Prorocentrum* including the description of *Prorocentrum donghaiense* Lu sp. nov. from the East China Sea. *Chinese Journal of Oceanology and Limnology, 19*, 337–344.

Lu, S., & Hodgkiss, I. J. (2004). Harmful algal bloom causative collected from Hong Kong waters. *Hydrobiologia, 512*, 231–238.

Luqman, M., Javed, M. M., Yousafzai, A., Saeed, M., Ahmad, J., & Chaghtai, F. (2015). Blooms of pollution indicator micro-alga (*Synedra acus*) in northern Arabian Sea along Karachi, Pakistan. *Indian Journal of Geo-marine Sciences, 44*(9), 1377–1381.

Mackenzie, F. T., De Carlo, E. H., & Lerman, A. (2011). Coupled C, N, P, and O biogeochemical cycling at the Land–Ocean interface. In E. Wolanski & D. S. McLusky (Eds.), *Treatise on estuarine and coastal science* (Vol. 5, pp. 317–342). Waltham: Academic.

Madhu, N. V., Jyothibabu, R., Maheswaran, P. A., Gerson, V. J., Gopalakrishnan, T. C., & Nair, K. K. C. (2006). Lack of seasonality in phytoplankton standing stock (chlorophyll-*a*) and production in western Bay of Bengal. *Continental Shelf Research, 26*, 1868–1883.

Madhu, N. V., Reny, P. D., Paul, M., Ullas, N., & Resmi, P. (2011). Occurrence of red tide caused by *Karenia mikimotoi* (toxic dinoflagellate) in the Southwest coast of India. *Indian Journal of Geo-marine Sciences, 40*(6), 821–825.

Margolis, L. (1993). A multi-species plankton bloom in Departure Bay. *Aquaculture Update, 62*, 1–3.

Mashiatullah, A., Qureshi, R. M., Ahmad, N., Khalid, F., & Javed, T. (2009). Physico-chemical and biological water quality of Karachi coastal water. *The Nucleus, 46*(9), 53–59.

Matrai, P. A., & Keller, M. D. (1994). Total organic sulfur and dimethylsulfoniopropionate in marine phytoplankton: Intracellular variations. *Marine Biology, 119*(1), 61–68.

Matsusato, T., & Kobayashi, H. (1974). Studies on the death of fish caused by red tide. *Bulletin of the Nansei Regional Fisheries Research Laboratory, 7*, 43–67.

Meng, P. J., Lee, H. J., Tew, K. S., & Chen, C. C. (2015). Effect of a rainfall pulse on phytoplankton bloom succession in a hyper-eutrophic subtropical lagoon. *Marine and Freshwater Research, 66*, 60–69.

Meng, P. J., Tew, K. S., Hsieh, H. Y., & Chen, C. C. (2016). Relationship between magnitude of phytoplankton blooms and rainfall in a hyper-eutrophic lagoon: A continuous monitoring approach. *Marine Pollution Bulletin*. https://doi.org/10.1016/j.marpolbul.2016.12.040.

Meng, P., Tew, K. S., Hsieh, H., & Chen, C. (2017). Relationship between magnitude of phytoplankton blooms and rainfall in a hyper-eutrophic lagoon: A continuous monitoring approach. *Marine Pollution Bulletin, 124*, 897–902.

Milly, P. C. D., Wetherald, R. T., Dunne, K. A., & Delworth, T. L. (2002). Increasing risk of great floods in a changing climate. *Nature, 415*, 514–517.

Mishra, S., & Panigraphy, R. C. (1995). Occurrence of diatom blooms in Bahuda estuary, East Coast of India. *Indian Journal of Marine Science, 24*, 99–101.

Mishra, S., Sahu, G., Mohanty, A. K., Singh, S. K., & Panigrahy, R. C. (2006). Impact of the diatom *Asterionella glacialis* (Castracane) bloom on the water quality and phytoplankton community structure in coastal waters of Gopalpur Sea, Bay of Bengal. *Asian Journal of Water, Environment and Pollution, 3*(2), 71–77.

Mohanty, A. K., Satpathy, K. K., Sahu, G., Hussain, K. J., Prasad, M. K. V., & Sarkar, S. K. (2010). Bloom of *Trichodesmium erythraeum* (Ehr.) and its impact on water quality and plankton community structure in the coastal waters of southeast coast of India. *Indian Journal of Marine Science, 39*(3), 323–333.

Munshi, A. B., Hina, A. S., & Usmani, T. H. (2005). Determination of levels of PCBs in small fishes from three different coastal areas of Karachi, Pakistan. *Pakistan Journal of Science Industrial Research, 48*, 247–251.

Murrell, M. C., & Lores, E. M. (2004). Phytoplankton and zooplankton seasonal dynamics in a subtropical estuary: Importance of cyanobacteria. *Journal of Plankton Research, 26*, 71–382.

Mutshinda, C. M., Finkel, Z. V., & Irwin, A. J. (2013). Which environmental factors control phytoplankton populations? A Bayesian variable selection approach. *Ecological Modelling, 269*, 1–8.

Nagabhushanam, A. K. (1967). On an unusually dense phytoplankton bloom around Minicoy Island (Arabian Sea) and its effect on tuna fisheries. *Current Science, 36*, 611.

Nair, V. R. (2013). *Status of flora and fauna of Gulf of Kachchh* (Vol. 87, p. 157). Goa: National Institute of Oceanography.

Nair, V. R., Devasssy, V. P., & Qasim, S. Z. (1981). Zooplankton and *Trichodesmium phenomenon*. *Indian Journal of Marine Science, 9*, 1–6.

Naqvi, S. W. A., George, M. D., Narvekar, P. V., Jayakumar, D. A., Shailaja, M. S., Sardesai, S., et al. (1998). Severe fish mortality associated with 'red tide' observed in the sea off Cochin. *Current Science, 75*, 543–544.

Narayana, S., Chitra, J., Tapase, S. R., Thamke, V., Karthick, P., Ramesh, C., et al. (2014). Toxicity studies of *Trichodesmium erythraeum* (Ehrenberg, 1830) bloom extracts, from Phoenix Bay, Port Blair, Andamans. *Harmful Algae, 40*, 34–39.

Naz, T., Burhan, Z., Munir, S., & Siddiqui, P. J. A. (2012). Taxonomy and seasonal distribution of *Pseudonitzschia* species (Bacillariophyceae) from the coastal water of Pakistan. *Pakistan Journal of Botany, 44*(4), 1467–1473.

Nergis, Y., Sharif, M., Farooq, M. A., Hussain, A., & Butt, J. A. (2012). Impact of industrial and sewage effluents on Karachi coastal water and sediment quality. *Middle East Journal of Scientific Research, 11*, 1443–1454.

Nishikawa, T., Hori, Y., Nagai, S., Miyahara, K., Nakamura, Y., Harada, K., et al. (2011). Long time-series observations in population dynamics of the harmful diatom *Eucampia zodiacus* and environmental factors in Harima-Nada, eastern Seto Inland Sea, Japan during 1974–2008. *Plankton and Benthos Research, 6*(1), 26–34.

Nishikawa, T., Hori, Y., Nagai, S., Miyahara, K., Nakamura, Y., Harada, K., et al. (2014). Long-term (36-year) observations on the dynamics of the fish-killing raphidophyte *Chattonella* in Harima-Nada, eastern Seto Inland Sea, Japan. *Journal of Oceanography, 70*(2), 153–164.

Onoue, Y., & Nozawa, K. (1989). Separation of toxin from harmful red tides occurring along the coast of Kogoshima prefecture. In T. Okaichi, D. M. Anderson, & T. Nemoto (Eds.), *Red tides: Biology, environmental science, and technology* (pp. 371–374). New York: Elsevier Science.

Padmakumar, K. B., Menon, N. R., & Sanjeevan, V. N. (2012). Is occurrence of harmful algal blooms in the exclusive economic zone of India on the rise? *International Journal of Oceanography, 2012*, 1–7.

Palmer, T. N., & Ralsanen, J. (2002). Quantifying the risk of extreme seasonal precipitation events in a changing climate. *Nature, 415*, 512–514.

Pant, A., & Devassy, V. P. (1976). Release of extracellular matter during photosynthesis by a *Trichodesmium* bloom. *Current Science, 45*, 487–489.

Piepenburg, D., Voss, J., & Gutt, J. (1997). Assemblages of sea stars (Echinodermata: Asteroidea) and brittle stars (Echinodermata: Ophiuroidea) in the Weddell Sea Antarctica and off Northeast Greenland(artic): A comparison of diversity and abundance. *Polar Biology, 17*, 305–322.

Power, S., Delage, F., Chung, C., Kociuba, G., & Keay, K. (2013). Robust twenty-first-century projections of El Nino and related precipitation variability. *Nature, 502*, 541–545.

Prabhu, M. S., Ramamurthy, S., Kuthalingam, M. D. K., & Dhulkheid, M. H. (1965). On an unusual swarming of the planktonic blue green algae *Trichodesmium* Spp. off Mangalore. *Current Science, 34*, 95.

Prabhu, M. S., Ramamurthy, S., Dhulkhed, M. H., & Radhakrishnan, N. S. (1971). *Trichodesmium* bloom and failure of oil sardine fishery. *Mahasagar, 4*, 62.

Prasath, B., Nandakumar, R., Jayalakshmi, T., & Santhanam, P. (2014). First report on the intense cyanobacteria *Microcystis aeruginosa* Kützing, 1846 bloom at Muttukadu Backwater, southeast coast of India. *Indian Journal of Geo-marine Sciences, 43*(2), 258–262.

Price, A. R. G., Keeling, M. J., & O'Calllaghan, C. J. (1999). Ocean-scale patterns of biodiversity of Atlantic asteroids determined from taxonomic distinctness and other measures. *Biological Journal of the Linnean Society, 66*, 187–203.

Qadri, M., Nergis, Y., Mughal, N. A., Sharif, M., & Farooq, M. A. (2011). Impact of marine pollution at Karachi coast in perspective of Lyari river. *American-Eurasian Journal of Agricultural & Environmental Sciences, 10*, 737–743.

Qasim, S. Z. (1970). Some characteristic of a *Trichodesmium* bloom in the Laccadives. *Deep Sea Research, 17*, 655–660.

Quinn, P. K., & Bates, T. S. (2011). The case against climate regulation via oceanic phytoplankton sulphur emissions. *Nature, 480*(7375), 51–56.

Qureshi, S. M., Mashiatullah, A., Rizvi, S. H. N., Khan, S. H., Javed, T., & Tasneem, M. A. (2001). Marine pollution studies in Pakistan by nuclear technology. *The Nucleus, 38*, 41–51.

Rabbani, M. M., Rehman, A. U., & Harms, C. E. (1990). Mass mortality of fishes caused by dinoflagellate bloom in Gwadar Bay, southwestern Pakistan. In: E. Graneli, B. Sundstroem, L. Edler & D. M. Anderson (Eds.), *Toxic marine phytoplankton* (pp. 209–214). Karachi: National Institute of Oceanography.

Raghuprasad, R., & Jayaraman, R. (1954). Preliminary studies on certain changes in the plankton and hydrological conditions associated with the swarming of *Noctiluca*. *Proceedings of the Indian Academy of Sciences, 40*, 49–57.

Rajagopalan, M. (2007). *Trichodesmium* (Ehr.) bloom along the southwest coast of India (Arabian Sea) and its impact on trace metal concentrations in seawater. *Estuarine, Coastal and Shelf Science, 71*, 641–646.

Ramamurthy, V. D., Selva Kumar, R. A., & Bhargava, R. M. S. (1972). Studies on the blooms of Trichodesmium erythraeum (EHR) in the waters of the Central west coast of India. *Current Science, 41*, 803–805.

Raymont, J. E. G. (1980). *Plankton and productivity in the oceans. Part. I. Phytoplankton* (p. 489). Oxford: Pergamon Press.

Reed, R. H., & Stewart, W. D. P. (1988). The responses of cyanobacteria to salt stress. In L. J. Rogers & J. R. Gallon (Eds.), *Biochemistry of the algae and cyanobacteria* (Vol. 12, pp. 217–231). Oxford: Clarendon Press.

Reginald, M. (2007). Studies on the importance of micro algae in solar salt production. *Seaweed Research Utilization, 29*, 151–184.

Reynolds, C. S. (2006). *The ecology of phytoplankton* (p. 402). Cambridge: Cambridge University Press.

Reynolds, C. S., Jaworski, G. H. M., Cmiech, H. A., & Leedale, G. F. (1981). On the annual cycle of the blue – Green algae *Microcystis aeruginosa* Kutz. Emend. Elenkin. *Philosophical Transactions of The Royal Society B Biological Sciences, 293*, 419–477.

Rogers, K., Clarke, K. R., & Reynolds, J. D. (1999). The taxonomic distinctness of coastal bottom-dwelling fish communities of the North-East Atlantic. *The Journal of Animal Ecology, 68*, 769–782.

Sachithanandam, V., Mohan, P. M., Karthik, R., Elangovan, S. S., & Padmavathi, G. (2013). Climate change influence the phytoplankton bloom (prymnesiophyceae: *Phaeocystis* spp.) in North Andaman coastal region. *Indian Journal of Geo-marine Sciences, 42*, 58–66.

Sahayak, S., Jyothibabu, R., Jayalakshmi, K. J., Habeebrehman, H., Sabu, P., Prabhakaran, M. P., Jasmine, P., Shaiju, P., George, R. M., Thresiamma, J., & Nair, K. K. C. (2005). Red tide of *Noctiluca miliaris* off south of Thiruvananthapuram subsequent to the 'stench event' at the southern Kerala coast. *Current Science, 89*, 1472–1473.

Sahu, B. K., Begum, M., Khadanga, M., Jha, D. K., Vinithkumar, N., & Kirubagaran, R. (2013). Evaluation of significant sources influencing the variation of physico-chemical parameters in Port Blair Bay, South Andaman, India by using multivariate statistics. *Marine Pollution Bulletin, 66*, 246–251.

Sahu, B. K., Begum, M., Kumarasamy, P., Vinithkumar, N. V., & Kirubagaran, R. (2014a). Dominance of *Trichodesmium* and associated biological and physico-chemical parameters in coastal water of Port Blair, South Andaman Island. *Indian Journal of Geo-Marine Sciences, 43*, 1739–1745.

Sahu, B. K., Begum, M., Kumarasamy, P., Vinithkumar, N., & Kirubagaran, R. (2014b). Dominance of *Trichodesmium* and associated biological and physico-chemical parameters in coastal waters of Port Blair, South Andaman Island. *Indian Journal of Geo-marine Sciences, 43*, 1–7.

Sahu, G., Mohanty, A. K., Acharya, M. S., Sarkar, S. K., & Satpathy, K. K. (2015). Changes in mesozooplankton community structure during *Trichodesmium erythraeum* bloom in the coastal waters of southwestern Bay of Bengal. *Indian Journal of Geo-marine Sciences, 44*(9), 1282–1293.

Saifullah, S. M., Khan, S. H., & Iftikhar, S. (2000). *Distribution of a trace metal Iron in mangrove habitat of Karachi*. Symposium on Arabian Sea as a Resource of Biological diversity, Pakistan.

Saifullah, S. M., Ismail, S., & Khan, S. H. (2002a). Copper contamination in Indus delta mangrove of Karachi. In *Prospectus for saline agriculture* (p. 447). Cham: Springer.

Saifullah, S. M., Khan, S. H., & Ismail, S. (2002b). Distribution of nickel in a polluted mangrove habitat of the Indus Delta. *Marine Pollution Bulletin, 44*, 570–576.

Santhanam, R. (1976). PhD thesis. Annamalai university, Chidambaram, India, p. 101.

Santhanam, R., Srinivasan, A., Ramadhas, V. M., & Devraj, P. (1994a). Impact of *Trichodesmium* bloom on the plankton and productivity in the tuticorin bay, southeast coast of India. *Indian Journal of Marine Science, 23*, 27–30.

Santhanam, R., Srinivasan, A., Ramadhas, V., & Devaraj, M. (1994b). Impact of *Trichodesmium* bloom on the plankton and productivity in the Tuticorin Bay, southeast coast of India, *Indian. Journal of Marine Science, 23*, 27–30.

Santhanam, P., Balaji Prasath, B., Nandakumar, R., Jothiraj, K., Dinesh Kumar, S., Ananth, S., Prem Kumar, C., Shenbaga Devi, A., & Jayalakshmi, T. (2013). Bloom in the Muthupettai mangrove lagoon, Southeast coast of India. *Seaweed Research Utilization, 35*, 178–186.

Sarangi, R. K., Prakash, C., & Nayak, S. R. (2004). Detection and monitoring of Trichodesmium bloom in the coastal waters of Sourashtra coast, India using IRS P4 OCM data. *Current Science, 86*, 1636–1841.

Sarangi, R. K., Chauhan, P., & Nayak, S. R. (2005). Inter-annual variability of phytoplankton blooms in the northern Arabian Sea during winter monsoon period (February–March) using IRS-P4 OCM data. *Indian Journal of Marine Sciences, 34*(2), 163–173.

Sargunam, C. A., Rao, V. N. R., & Nair, K. V. K. (1989). Occurrence of *Noctiluca* bloom in Kalpakkam coastal waters, east coast of India. *Indian Journal of Marine Science, 18*, 289–290.

Sarkar, S. K., Saha, M., Takada, H., Bhattacharya, A., Mishra, P., & Bhattacharya, B. (2007). Water quality management in the lower stretch of the river Ganges, east coast of India: An approach through environmental education. *Journal of Cleaner Production, 15*(16), 1559–1567.

Sasmal, S. K., Panigrahy, R. C., & Mishra, S. (2005). *Asterionella* blooms in the northwestern Bay of Bengal during, 2004. *International Journal of Remote Sensing, 26*(10), 3853.

Satpathy, K. K., & Nair, K. V. K. (1996). Occurrence of phytoplankton bloom and its effect on coastal water quality. *Indian Journal of Marine Science, 25*, 145–147.

Satpathy, K. K., Mohanty, A. K., Sahu, G., Prasad, M. V. R., Venkatesan, R., Natesan, U., & Rajan, M. (2007). On the occurrence of *Trichodesmium erythraeum* (Ehr.) bloom in the coastal waters of Kalpakkam, east coast of India. *Indian Journal of Science and Technology, 1*(2), 1–9.

Saunders, R. D., & Glenn, D. A. (1969). Diatoms. *Memoirs of the Hourglass Cruises, 1*(3), 1–119.

Savage, R. E., & Wimpenny, R. S. (1936). Phytoplankton and the Herring, Part II 1933–1934. *Ministry of Agriculture and Fisheries Investments*, Series II 1936, *15*(1), 1–88.

Selvakumar, K., & Sundararaman, M. (2007). Diversity of cyanobacterial flora in the backwaters of Palk bay region. *Seaweed Research Utilization, 29*, 139–144.

Shetty, H. P. C., & Saha, S. B. (1971). On the significance of the occurrence of blooms of the diatom *Hemidiscus hardmannianus* (Greville) Mann in relation to Hilsa fishery in Bengal. *Current Science, 40*(15), 410–411.

Shetty, H. P. C., Gupta. T. R. C., & Kattai, R. J. (1988). *Green water phenomena in the Arabian Sea off Mangalore*. In Proceedings of the first India fisheries forum, pp. 339–346.

Shumway, E. S. (1990). A review of the effects of algal blooms on shellfish and aquaculture. *Journal of the World Aquaculture Society, 21*, 65–105.

Sivonen, K. (1996). Cyanobacterial toxins and toxin production. *Phycologia, 35*, 12–24.

Six, K. D., Kloster, S., Ilyina, T., Archer, S. D., Zhang, K., & Maier-Reimer, E. (2013). Global warming amplified by reduced sulphur fluxes as a result of ocean acidification. *Nature Climate Change, 3*, 975–978.

Smayda, T. J. (1997). Bloom dynamics: Physiology, behavior, trophic effects. *Limnology and Oceanography, 42*(5 part 2), 1132–1136.

Smayda, T. J. (2007). Reflections on the ballast water dispersal- harmful algal bloom paradigm. *Harmful Algae, 6*, 601–622.

Staehr, P. A., Testa, J. M., Kemp, W. M., Cole, J. J., Sand-Jensen, K., & Smith, S. V. (2012). The metabolism of aquatic ecosystems: History, applications, and future challenges. *Aquatic Sciences, 74*, 15–29.

Stefels, J., Steinke, M., Turner, S., Malin, G., & Belviso, S. (2007). Environmental constraints on the production and removal of the climatically active gas dimethylsulphide (DMS) and implications for ecosystem modelling. *Biogeochemistry, 83*(1–3), 245–275.

Strickland, J. R. D., & Parsons, T. R. (1972). *A practical handbook of seawater analysis* (Bulletin 167, pp. 310). Ottawa: Fisheries Research Board of Canada.

Subba Rao, S. D. V. (1969). *Asterionella japonica* bloom and discolouration off Waltair, Bay of Bengal. *Limnology and Oceanography, 14*, 632–634.

Subba Rao, D. V., Pan, Y., & Smith, S. J. (1995). Allelopathy between *Rhizosolenia alata* (Brightwell) and the toxigenic *PseudoNitzschia pungens* f. multiseries (Hasle). In P. Lassus, G. Arzul, E. Erard, P. Gentien, & C. Marcaillou (Eds.), *Harmful marine algal blooms* (pp. 681–686). Paris: Lavoisier Intercept.

Subrahmanyan, R. (1954). On the life history and ecology of *Hornellia marina* Gen. Ersp. Nov, (Chloromonodinaeae), causing green discolouration of the sea and mortality among marine organisms off the Malabar coast. *Proceedings of the Indiana Academy of Sciences, 39*, 182–203.

Subrahmanyan, R. (1968). *The Dinophyceae of the Indian seas. Part 1. Genus Ceratium* (p. 129). Cochin: The City Press.

Subrahmanyan, R. (1971). *The Dinophyceae of the Indian seas. Part 2. Genus Peridinium* (p. 334). Cochin: The City Press.

Subrahmnayan, R. (1959). Studies on the phytoplankton of the west coast of India. Part I. Quantitative and qualitative fluctuation of total phytoplankton crop, the zooplankton crop and their interrelationship with remarks on the magnitude of the standing crop and production of matter and their relationship to fish landings. *Proceedings of the Indiana Academy of Sciences, 50*, 113–187.

Subramanian, A., & Purushothaman, A. (1985). Mass mortality of fish and invertebrates associated with a bloom of Hemidiscus hardmannianus (Bacillariophyceae) in Parangipettai (Southern India). *Limnology and Oceanography, 30*(4), 910–911.

Suikkanen, S., Fistarol, G. O., & Granéli, E. (2005). Effects of cyanobacterial allelochemicals on a natural plankton community. *Marine Ecology Progress Series, 287*, 1–9.

Sweeney, B. M. (1976). *Pedinomonas noctilucae* (Prasinophyceae) the flagellate symbiotic in *Noctiluca* (Dinophyceae) in Southeast Asia. *Journal of Phycology, 12*, 460–464.

Takahashi, M., Seibert, D. L., & Thomas, W. H. (1977). Occasional blooms of phytoplankton during summer in Saanich Inlet, BC, Canada. *Deep Sea Research, 24*(8), 775–780.

Takayama, H., & Adachi, R. (1984). *Gymnodinium nagasakiense* sp nov., a red-tide forming dinophyte in the adjacent waters of the Sea of Japan. *Bulletin of the Plankton Society of Japan, 31*, 7–14.

Tan, J., Jakob, C., Rossow, W. B., & Tselioudis, G. (2015). Increases in tropical rainfall driven by changes in frequency of organized deep convection. *Nature, 519*, 451–454.

Tanaka, K., Muto, Y., & Shimada, M. (1994). Generation of superoxide anion radicals by the marine phytoplankton organism *Chattonella antique*. *Journal of Plankton Research, 16*, 161–169.

Tang, D. L., Di, B. P., Wei, G., Ni, I., Oh, I. S., & Wang, S. (2006). Spatial, seasonal and species variations of harmful algal blooms in the South Yellow Sea and East China Sea. *Hydrobiologia, 568*, 245–253.

Tangen, K. (1977). Blooms of *Gyrodinium aureolum* (Dinophyceae) in North European waters accompanied by mortality of marine organisms. *Sarsia, 63*, 123–133.

Thajuddin, N., Nagasathya, A., Chelladevi, R., & Saravanan, I. (2002). Biodiversity of cyanobacteria in different salt pans of Pudukkottai District, Tamilnadu. *Seaweed Research and Utilization, 24*, 1–11.

Tillmann, U., & John, U. (2002). Toxic effects of *Alexandrium* spp. on heterotrophic dinoflagellates: An allelochemical defence mechanism independent of PSP-toxin content. *Marine Ecology Progress Series, 230*, 47–58.

Tomas, C. R. (1996). *Identifying marine diatoms and dinoflagellates* (p. 598). New York: Academic.

Tomas, C. R. (1997). *Identifying marine phytoplankton*. Academic Press, USA. UNESCO. Protocols for the joint global ocean flux study (JGOFS). Manual and guides (Vol. 29, p. 170).

Uher, G. (2006). Distribution and air-sea exchange of reduced sulphur gases in European coastal waters. *Estuarine Coastal and Shelf Science, 70*, 338–360.

Ulrich, R. M., Casper, E. T., Campbell, L., Richardson, B., Heil, C. A., & Paul, J. H. (2010). Detection and quantification of *Karenia mikimotoi* using real-time nucleic acid sequence-based amplification with internal control RNA (IC-NASBA). *Harmful Algae, 9*, 116–122.

Vallina, S. M., Simó, R., & Manizza, M. (2007). Weak response of oceanic dimethylsulfide to upper mixing shoaling induced by global warming. *Proceedings of the National Academy of Sciences of the United States of America, 104*, 16004–16009.

Velankar, A. D., & Chaugule, B. B. (2007). Algae of the salt pans of Nalasopara, Mumbai. *Seaweed Research Utilization, 29*, 273–278.

Verity, P. G., & Villareal, T. A. (1986). The relative food value of diatoms, dinoflagellates, flagellates and cyanobacteria for tintinnid ciliates. *Archiv für Protistenkunde, 31*, 71–84.

Wang, S., Tang, D. L., He, F. L., & Aza, Y. F. (2008). Occurrences of harmful algal blooms (HABs) associated with ocean environments in the South China Sea. *Hydrobiologia, 596*, 79–93.

Westberry, T. K., & Siegel, D. A. (2006). Spatial and temporal distribution of *Trichodesmium* in the world's oceans. *Global Biogeochemical Cycles, 20*, GB4016.

White, W. A. (1981). Marine zooplankton can accumulate and retain dinoflagellate toxins and cause fish kills. *Limnology and Oceanography, 28*, 103–109.

Wolfe, J. M., & Rice, E. L. (1979). Allelopathic interactions among algae. *Journal of Chemical Ecology, 5*(4), 533–542.

Yamamoto, T. (2003). The Seto Inland Sea – Eutrophic or oligotrophic? *Marine Pollution Bulletin, 47*(1), 37–42.

Yoshinaga, I., Hitomi, T., Miura, A., Shiratani, E., & Miyazaki, T. (2006). Cyanobacterium *Microcystis* bloom in a eutrophicated regulating reservoir. *Japan Agricultural Research Quarterly, 40*(3), 283–289.

Zhang, F., Ma, L., Xu, Z., Zheng, J., Shi, Y., Lu, Y., & Miao, Y. (2009). Sensitive and rapid detection of *Karenia mikimotoi* (Dinophyceae) by loop-mediated isothermal amplification. *Harmful Algae, 8*, 839–842.

Zhou, Z. X., Yu, R., & Zhou, M. J. (2017). Seasonal succession of microalgal blooms from diatoms to dinoflagellates in the East China Sea: A numerical simulation study. *Ecological Modelling, 360*(2017), 150–162.

Chapter 3
Harmful Algal Blooms (HABs)

Abstract Harmful algal blooms (HABs) are serious biological nuisances and become a global epidemic. This is primarily flagellate events, causing mass mortality, physiological impairment or other negative in situ effects. HABs are increasing their frequency, persistence, regional coverage/spatial extent and economic impact worldwide in recent decades as a result of changes in enhanced coastal eutrophication and climate change along with invasion of alien species through ballast water exchange. This also happens on account of the world's increasing trend of unscientific and irrational exploitation on the coastal zone for shelter, food, construction, food, aquaculture, recreation and other commercial uses, which results in increasing eutrophication. Naturally occurring red tides and harmful algal blooms (HABs) are of increasing importance in the eutrophic coastal environment and can have remarkable adverse impacts on coastal benthic and epipelagic communities. Determining the key regulatory factors is somehow problematic since algal blooms are often unpredictable, irregular or of short duration. The chapter has addressed an illustrative account of all the six categories of HABs along with description of the causative agents and their associated clinical symptoms. The recent tools and techniques developed towards operational status for prediction or detection of HABs have described. Finally, climate change impact on the occurrence of algal blooms in a global scale has also been illustrated.

Keywords Harmful Algal blooms (HABs) · PSP · DSP · ASP · NSP · CFP · AZA · CyanoHAB · Climate change

3.1 Introduction

3.1.1 Dinoflagellates: A Potent Agent for HABs

Dinoflagellates (class Dinophyceae) constitute one of the key components of the marine phytoplankton community, and they play an important role in marine food webs. This dominant member of tropical and subtropical phytoplankton displays a wide diversity of trophic strategies, including autotrophy, heterotrophy and mixotrophy and can migrate through the water column due to their motility. They possess

S. K. Sarkar, *Marine Algal Bloom: Characteristics, Causes and Climate Change Impacts*, https://doi.org/10.1007/978-981-10-8261-0_3

unique capacity to move vertically several metres per day and to cross a strong thermocline region. This gives them a superb advantage, because they can spend the daylight hours near the photic zone with high illumination but can also migrate at night to more nutrient-rich (eutrophic) water below. Dinoflagellates accomplish this by flagella movement, whereas cyanobacteria adjust their depth by changing cell buoyancy. This advantage can be explored exclusively a stable water column. Nuisance blooms, therefore, do not usually develop unless the surface part of the water column is stable or if the phytoplankton population is trapped near a discontinuity of water quality. These are usually become abundant in summer and autumn following diatom bloom, as they are better adopted while living in nutrient-impoverished water. For the same reason, dinoflagellates are usually dominated in the phytoplankton community in nutrient-poor tropical and subtropical waters and thus act as a potential causative agent for algal bloom in comparison to other microalgae. The availability and transport of a seabed pool of cysts may be significant in the origin of dinoflagellate red tides.

Dinoflagellate blooms can appear suddenly but often collapse within few days, when the cells can sink from the water column, generally aggregate or floc and finally sediment on the sea floor. Nutrients act as a 'limiting factor' for sustenance of microalgae which are exhausted due to their mass decay. Due to subsequent bacterial decomposition of organic matter standing crop or biomass, it depletes the available oxygen resulting hypoxic (<2 ml l^{-1} dissolved oxygen) or anoxic condition, resulting fish mortality from asphyxiation. Such oxygen depletion is the ubiquitous features for all algal blooms including dinoflagellates.

3.1.2 Red Tide and HABs

Red tide blooms have become fairly frequent resulting in discolouration of seawater (amber, red, reddish-brown, brown, yellow orange to purple) caused by the highly dense population of dinoflagellates in coastal regions (as shown in Fig. 3.1) – the best known source of dense and toxic bloom. The biotoxins are greatly harmful to resource species which may accumulate the toxins and transfer them to people. Accumulation and depuration rates vary with species, season, tissue and location, and many species are negatively affected by the toxin. This biological nuisance causes substantial disturbance of entire pelagic food web structure and functioning, along with dramatic alterations in phytoplankton community and recurrent hypoxic conditions. The low oxygen or hypoxic condition is due to high amounts of oxygen required for decomposition of the algae at the end of the bottom, causing asphyxiation to the pelagic organisms. On the other hand, deposition of dead microalgae might be a potential source of digestible particulate organic matter (POM) to benthic deposit feeders.

Red tides seem to be associated with sudden influxes of inorganic nutrients due to anthropogenic nutrient enrichment of inorganic nutrients such as phosphate, silicate and nitrate (eutrophication) as a result of increased discharges of industrial and

Fig. 3.1 *Noctiluca* bloom in New Zealand coastal waters. (Courtesy: Dr. Mindy L. Richlen, National Office for HABs, WHOI, USA)

urban waste waters, vitamins and humic substances into the coast from land sources. Chronic and episodic nutrient influxes, as a whole, promote HAB development. The accumulation of total organic carbon (TOC), total nitrogen (TN), total phosphorous (TP) and biogenic silica (BSi) in sediments is regarded as signals of eutrophication and diatom productivity (Wang et al. 2015). However, the nature of signals was different for eutrophication in different coastal regions in the world. It is worthwhile to mention that not all red tides are toxic and not all red water is caused by dinoflagellates. The Red Sea received its name because of dense blooms of non-toxic cyanobacteria with large amounts of red pigments. The Indian Ocean has red tides due to the presence of a toxic cyanobacterium, not a dinoflagellate.

Major associated factors triggering red tide events include warm ocean surface temperature, calm seas, disturbance of dormant dinoflagellates, low salinity along with heavy rain followed by sunny days and decreased natural filtering due to deforestation. Red tide has also occurred in the temperate and tropical waters of Japan, the Philippines, eastern Australia, Malaysia (particularly in Sabah), the United States, Peru, Chile, South Africa, England, Scandinavia and Papua New Guinea. Red tides have disastrous and diverse environmental hazards, such as mass mortality of fish, others contaminate shellfish and still others produce no toxic effects. Blooms may produce biotoxins, cause hypoxia and alter food webs; they can constitute an environmental health hazard, degrade water quality and habitat and are therefore deemed 'harmful algal blooms' (HABs), and the key species belong to diverse phytoplankton groups, such as diatoms, dinoflagellates, cyanobacteria and prasinophytes.

So far HAB occurred in China in 2013 is recognized as the largest one, stretching an area of 7500 square miles (https://phys.org/news/2013-07-china-largest-ever-algae-bloom.html), followed by another in 2015 which blanketed an even greater 13,500 square miles (http://www.ibtimes.co.uk). Heavy nutrient inputs to coastal waters from anthropogenic sources, large-scale, high-biomass blooms (those covering more than 1000 km^2) have been increasing exponentially along the coast (Shen et al. 2012). The relatively coarse spatial resolution of SeaWiFS data limits the study area of HABs to a large spatial scale (>1000 km^2) which is often found in coastal regions such as the East China Sea, Bohai, the Gulf of Mexico and the Korean coastal waters (Ahn and Shanmugam 2006).

HAB dynamics are determined by a complex interplay of abiotic and biotic factors, and their emergence has often been linked to eutrophication and more recently to climate change. HABs represent a major environmental problem, usually characterized by the proliferation and occasional dominance of a particular species of potent natural toxins or otherwise harmful alga. The changes of the dominant HAB-causing agents may have affected the timing of HAB occurrence in the Chinese coastal regions (Tang et al. 2006). Most interestingly, heterotrophic dinoflagellate *Noctiluca scintillans* (with bioluminescent capacity) and the centric diatom *Skeletonema costatum* were dominant causative species before 2000. Later on, the harmful planktonic dinoflagellate *Prorocentrum donghaiense* became dominant as a succession sequence during 2000–2004 causing large blooms (Tang et al. 2006). Such typical succession might be due to allelopathy, i.e. production of toxic compounds by a species to inhibit other competitors (as discussed in Chap. 2 under Sect. 2.2.3). Harmful algae are known to utilize this chemically mediated mechanism as a form of interference competition for limiting resources. However, only a solitary case has been documented in favour of allelopathy where the potent karlotoxins have shown to inhibit competitor growth and immobilize prey with effects depending on sterol composition of target species (Sheng et al. 2010). In contrast, no other toxic compounds (e.g. saxitoxin, brevetoxin or domoic acid) have been identified as allelopathic agent. Poulin et al. (2018) have demonstrated the biological variability in allelopathic potency of five genetic strains of the red tide dinoflagellate *Karenia brevis* using nuclear magnetic resonance (NMR) spectroscopy.

It is worth to mention that dinoflagellates typically require more nutrients that they cannot manufacture themselves than diatoms which might support the succession in such phenomenon. In contrast, *Noctiluca scintillans* is the most potential red tide causative agent in many temperate and subtropical coastal regions and solely responsible for majority of red tides in Hong Kong coastal regions (such as Tolo Harbour, Port Shelter), situated on the south coast of China (Wong 1989; Yin 2003; Liu and Wong 2006), accounting for >31% of the algal blooms recorded during 1980–2014 (HKEPD 2014). It is worth to refer that the species is non-toxic in nature and less harmful than most of the HAB-causing species but results oxygen depletion leading substantial damage for other coexisting pelagic organisms. Recently, Zhang et al. (2017) asserted that optimum hydrological characteristics (especially temperature and water stability) along with a rich food supply were the prerequisites conditions for their bloom and their spatial distribution is mainly governed by physical (winds, tides and currents) and biological coupling.

HABs have substantially increased in intensity and frequency extending different geographical locales over the last three decades, causing considerable economic (e.g. commercial and recreational fisheries) and ecological losses (e.g. loss of biodiversity). The possible reasons for this increase HAB phenomenon are as follows:

1. Increased release of nutrients to coastal waters, i.e. coastal eutrophication; both chronic and episodic nutrient delivery promote HAB development.
2. Increase of arrival of new algal recruits, which has been exacerbated by the transport of organisms for mariculture.
3. Rapid development of industry and agriculture adjacent to coastal regions.
4. Increased cultivation and harvesting of shellfish.
5. As a result of changes in oceanic climate.
6. Increased and enhanced long-distance dispersal in ballast water (Landsberg 2002; Heisler et al. 2008; Rabalais et al. 2009).

An apparent shift of bloom-forming innocuous diatom species to toxic or noxious dinoflagellates is possible with potent toxins (Liu et al. 2016a, b) which might be caused due to the factors as referred above.

HABs can have adverse impact on the fauna from the micro-scale to the ecosystem scale (Landsberg 2002; Graneli and Turner 2008), and their effects are related to the species involved and prevailing environmental conditions (Landsberg 2002; Sunda et al. 2006). A range of negative effects on fauna is being recorded by several researchers, such as (1) clogging of gills (Graneli and Turner 2008); (2) poison by water-borne toxins and cellular damage caused by reactive oxygen species (Cembella et al. 2002; Kim et al. 2002); (3) localized anoxia (Anderson et al. 2002); (4) fertilization success and embryonic and larval development of the sea urchin *Lytechinus variegatus*, a key ecological herbivore, which are negatively affected by the benthic toxic dinoflagellate *Ostreopsis* cf. *ovata* even at low abundances (Neves et al. 2018); and (5) predation (Vogelbein et al. 2002). As a result of the variety of mechanisms, HABs seem to affect a much wider variety of taxa than other stresses/disturbances so far recorded in marine ecosystems (Pratchett et al. 2008; Bauman et al. 2010), as evidenced from the damage caused by HAB and differences in the magnitude of events.

During a HAB event, algal toxins (biotoxins) can accumulate in predators and organisms higher up the food web. Toxins may also be present in ambient waters, where wave action or human activities can create aerosols containing toxins and cellular debris. Animals, including humans, can thus be exposed to HAB-related toxins from a set of mediums, such as through consumption of contaminated seafood, have contact with contaminated water or inhale contaminated aerosols (Backer and McGillicuddy 2006). Winds can blow ashore sea spray containing contaminated dinoflagellate cells from nearshore bloom which may result allergy-like reactions like irritated eyes, running noses, coughs, etc. to the local coastal people.

Nutrient loadings in coastal waters can cause eutrophication, leading to changes in Chl a, primary production, macro- and microalgal biomass as well as benthos biomass, benthos community structure and benthic macrophytes, which in turn result in major changes in species composition, structure and function of marine communities over large areas (Cloern 2001; Islam and Tanaka 2004). Microalgal blooms are not only a consequence of aquatic pollution but also natural phenomena linked with the seasonal cycle of photosynthetic organisms in aquatic ecosystems (Berdalet et al. 2016; Islam and Tanaka 2004). The rise in toxic and harmful algae has adverse impacts on living marine resources and public health.

3.2 Synoptic Account of the Bloom-Forming Species

3.2.1 Azadinium spinosum *Elbrächter and Tillmann, 2009*

Azadinium spinosum is a autotrophic dinoflagellate species that produces azaspiracid (AZA) toxins (toxins associated with shellfish poisoning) (Salas et al. 2011), particularly AZA 1, AZA 2 and an isomer of AZA 2. It measures 12–16 μm in length and 7–11 μm wide and is a peridinin-containing photosynthetic dinoflagellate with a thin theca. Its large nucleus is spherical and presents posteriorly, whereas its single chloroplast is parietal and lobed and extends into the epi- and hyposome (Tillmann et al. 2009). This was first isolated from the North Sea that can be easily misidentified as *Heterocapsa* or *Scrippsiella*. Originally, AZA was thought to be produced by the heterotrophic dinoflagellate, *Protoperidinium crassipes*. The distribution of this new species is currently unknown but will likely mirror the distribution pattern of the syndrome (https://products.coastalscience.noaa.gov/pmn/_docs/Factsheets/Factsheet_Azadinium.pdf). The primary toxin is a lipophilic, polyether toxin called azaspiracid (AZA-1). The exact mode of action is not explored properly; in vivo studies have shown effects towards the intestinal tract, lymphoid tissues and immune system cells. In vitro studies revealed that AZA increases levels of cytosolic calcium and cAMP in lymphocytes while reducing F-actin levels in neuroblastoma cells. The precise mode of action(s) is currently under investigation but may be related to inhibition of cholesterol synthesis.

3.2.2 Alexandrium tamarense *(Lebour, 1925) Balech, 1995*

Alexandrium tamarense is a species of dinoflagellates known to produce the neurotoxin saxitoxin which causes the human illness clinically known as paralytic shellfish poisoning (PSP). Multiple species of phytoplankton are known to produce saxitoxin, including at least ten other species from the genus *Alexandrium*. Molecular research result shows that this species belongs to the *Alexandrium tamarense* complex (Atama complex, including *A. tamarense*, *Alexandrium fundyense*, *Alexandrium catenella*) and that none of the three original morphospecies designations forms monophyletic groups in the present SSU-based and previous LSU-based (Lilly et al. 2007) phylogenetic trees, i.e. these species designations are invalid (Miranda et al. 2012).

Because of the similar morphological resemblance with *A. tamarense*, *A. catenella* and *A. fundyense*, they are grouped together within *A. tamarense* species complex (Anderson et al. 2012). PSP occurrences by the toxic *A. tamarense* species complex were only known in Europe, North America and Japan in the 1970s (Dale and Yentsch 1978). Their distribution, however, expanded widely from the subtropical to the subarctic of the north hemisphere and into the temperate south hemisphere (Hallegraeff 1993; Lilly et al. 2007), and it is presumed that the distributions of HAB species and their abundances have also been affected due to recent climate change impact (Hallegraeff 2010).

Matsuoka et al. (2018) demonstrated the coincidence of paralytic shellfish poisoning (PSP) outbreaks caused by *Alexandrium tamarense* (Dinophyceae) and occurrence of devastating tsunami generated by the Great East Japan Earthquake, March 11, 2011. Vertical distributions of *Alexandrium* cysts in two sediment cores from the Ofunato Bay revealed that the sediments above ca. 25 cm were eroded, resuspended and redeposited due to this natural catastrophic event. The sediment cores included unusually abundant *Alexandrium* cysts due to the specific cyst-forming capability of this species mainly due to redeposition of older sediments by the tsunami. In a similar event, the first PSP incident on 1961 in Ofunato Bay probably resulted from huge blooms of *A. tamarense* initiating from resting cysts which were moved to the sea floor by deep erosion and resuspension of sediments involving ellipsoidal *Alexandrium* cysts filled with fresh protoplasm after the Chilean tsunami in 1960.

3.2.3 **Pseudo-nitzschia multiseries** ***(Hasle) Hasle, 1995***

Pseudo-nitzschia is a marine planktonic diatom genus containing some species capable of producing the neurotoxin domoic acid (DA), which is responsible for the neurological disorder known as amnesic shellfish poisoning. Currently, 48 species are known, 23 of which have been shown to produced DA. It was originally hypothesized that only dinoflagellates could produce harmful algal toxins, but a deadly bloom of *Pseudo-nitzschia* occurred in 1987 in the bays of Prince Edward Island, Canada, and led to an outbreak of ASP (Maldonado 2002). Blooms have since been characterized in coastal waters worldwide and have been linked to increasing marine nutrient concentrations (Parsons and Dortch 2002).

Pseudo-nitzschia species are bilaterally symmetrical pinnate. Cell walls are made up of elongated silica frustules. The silica wall is fairly dense which leads to negative buoyancy, providing a number of advantages. The wall allows the diatoms to sink to avoid light inhibition or nutrient limitations, as well as to protect against grazing zooplankton. The silica frustules also contribute vastly to the sediment layers of the earth and to the fossil record, which make them exceptionally useful in increasing understanding of numerous processes such as gauging the degree of climate change. Before sinking to the ocean floor, every atom of silicon that enters the ocean is integrated into the cell wall of a diatom about 40 times (Kuwata and Jewson 2015).

Silica frustules contain a central raphe, which secretes mucilage that allows the cells to move by gliding. Cells are often found in overlapped, stepped colonies and exhibit collective motility (Lundholm and Moestrup); *Pseudo-nitzschia* species synthesize their own food through the use of light and nutrients in photosynthesis. The diatoms have a central vacuole to store nutrients for later use and a light-harvesting system to protect them against high-intensity light.

Pseudo-nitzschia species is cosmopolitan in nature, documented along the Pacific coast from Canada to California, along the Atlantic Northeast coast of Canada, North Carolina and the Gulf of Mexico (NOAA Coastal Science). Recently, they have been detected in the open ocean as well as gulfs and bays, showing a presence in many diverse environments, including off the coasts of Canada, Portugal, France, Italy, Croatia, Greece, Ireland and Australia (Dhar et al. 2015). In general, diatoms flourish in turbulent, nutrient-rich waters.

3.2.4 **Noctiluca scintillans** *(Macartney) Kofoid and Swezy, 1921*

Noctiluca scintillans is a non-photosynthetic heterotrophic and phagotrophic dinoflagellate species; chloroplasts are absent and the cytoplasm is mostly colourless. The presence of photosynthetic symbionts can cause the cytoplasm to appear pink or green in colour (Sweeney 1978). A number of food vacuoles are present within the cytoplasm. A large eukaryotic nucleus is located near the ventral groove with cytoplasmic strands extending from it to the edge of the cell (Dodge 1982; Fukuyo et al. 1990; Hallegraeff 1991; Steidinger and Tangen 1996). This is a distinctively shaped athecate species in which the cell is not divided into epitheca and hypotheca. Cells are very large, inflated (balloon-like) and sub-spherical. The ventral groove is deep and wide and houses a flagellum, a tooth and a tentacle. Only one flagellum is present in this species and is equivalent to the transverse flagellum in other dinoflagellates. The tooth is a specialized extension of the cell wall. The prominent tentacle is striated and extends posteriorly. Cells have a wide range in size, from 200 to 2000 μm in diameter (Dodge 1982; Fukuyo et al. 1990; Hallegraeff 1991; Taylor et al. 1995; Steidinger and Tangen 1996).

Noctiluca scintillans is an unarmoured, marine, red tide-causing planktonic dinoflagellate species and bioluminescent in many temperate and subtropical coastal regions. This red tide heterotrophic species is cosmopolitan in nature and distributed in cold and warm waters and presents in two forms: (1) red *Noctiluca* is heterotrophic and acts as a microzooplankton grazer in the food web, and (2) green *Noctiluca* contains the photosynthetic symbiont *Pedinomonas noctilucae*

(Subrahmanyan) (prasinophyte), but it also feeds on other plankton when the food supply is abundant. Red *Noctiluca* has wide geographical distribution in the temperate to subtropical coastal regions of the world. It occurs over a wide temperature range of about 10–25 °C and at higher salinities (generally not in estuaries). It is particularly abundant in high biological productivity areas such as upwelling or eutrophic regions where diatoms dominate, since they are its preferred food source. Green *Noctiluca* is much more restricted to a temperature range of 25–30 °C and occurs mainly in the tropical waters of Southeast Asia, Bay of Bengal (east coast of India), in the eastern, western and northern Arabian Sea and the Red Sea and has become abundant in the Gulf of Oman. Red and green *Noctiluca* overlap in their distribution in the eastern, northern and western Arabian Sea with a seasonal shift from green *Noctiluca* in the cooler, winter convective mixing, higher productivity season to red *Noctiluca* in the more oligotrophic, warmer summer season (Dodge 1982; Fukuyo et al. 1990; Hallegraeff 1991; Taylor 1993; Taylor et al. 1995; Steidinger and Tangen 1996 Harrison et al. 2011).

The species reproduces both asexually (by binary fission) and sexually (isogamete formation). This species has a diplontic life cycle: the vegetative cell is diploid, while the gametes are haploid and gymnodinioid with dinokaryotic nuclei (Zingmark 1970). This is a strongly buoyant planktonic species common in neritic and coastal regions of the world. It is also bioluminescent in some parts of the world. This enormous bloom-forming species is associated with mass mortalities in fish and marine invertebrate events. *N. scintillans* red tides frequently form in spring to summer in many parts of the world often resulting in a strong pinkish red or orange discolouration of the water (tomato soup). Blooms have been reported from Australia (Hallegraeff 1991), Japan, Hong Kong and China (Huang and Qi 1997) where the water is discoloured red. The Turkish Straits System (TSS) includes the Bosphorus, Sea of Marmara and Dardanelles (Unsal et al. 2003; Turkoglu et al. 2004; Turkoglu and Buyukates 2005; Turkoglu 2010; Turkoglu and Erdogan 2010) and the Black Sea (Porumb 1992; Turkoglu and Koray 2002, 2004), where the water is discoloured red. However, not all blooms associated with *N. scintillans* are red. The colour of *N. scintillans* is in part derived from the pigments of organisms inside the vacuoles of *N. scintillans*. For instance, green tides result from *N. scintillans* populations containing green-pigmented prasinophytes (Chlorophyta) that are living in their vacuoles (Hausmann et al. 2003). Blooms in New Zealand were reported pink with cell concentrations as high as 1.9×10^6 cells/L (Chang 2000). In Indonesia, Malaysia and Thailand (tropical regions), however, the water colour is green due to the presence of green prasinophyte endosymbionts (Sweeney 1978; Dodge 1982; Fukuyo et al. 1990; Hallegraeff 1991; Taylor et al. 1995; Steidinger and Tangen 1996).

This large cosmopolitan species is distributed worldwide in cold and warm waters. Populations are commonly found in coastal areas and embayment of tropical and subtropical regions (Dodge 1982; Fukuyo et al. 1990; Hallegraeff 1991; Taylor et al. 1995; Steidinger and Tangen 1996). This species is frequently referred to as *N. miliaris*, although Macartney's specific name has priority. Taylor (1976) suggests that the simplest solution to the problem of nomenclature is to accept the priority of the 'scintillans' especially as this has been used by two major works

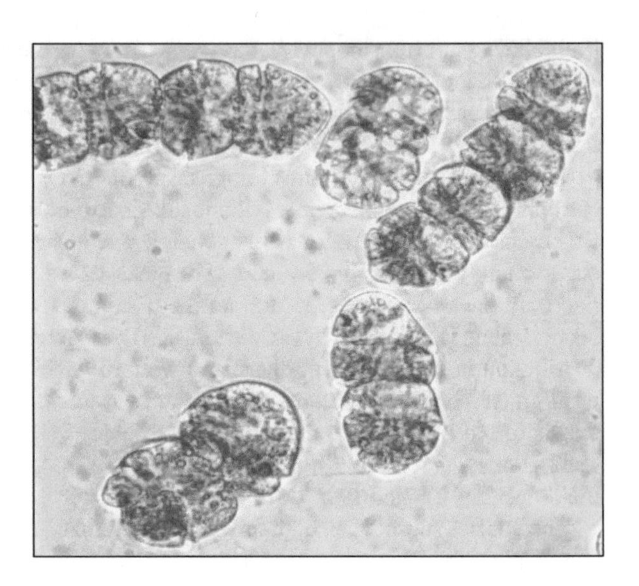

Fig. 3.2 Dinoflagellate *Cochlodinium polykrikoides* (Dinophyceae) recorded in Long Island, New York, USA. (Courtesy: Dr. Christopher J. Gobler, Stony Brook University)

(Kofoid and Swezy 1921; Lebour 1925). This is a phagotrophic species, feeding on phytoplankton (mainly diatoms and other dinoflagellates), protozoans, detritus and fish eggs (Dodge 1982; Fukuyo et al. 1990; Hallegraeff 1991; Taylor et al. 1995; Steidinger and Tangen 1996).

Toxic blooms of *N. scintillans* have been linked to massive fish and marine invertebrate kills. Although this species does not produce a toxin, it has been found to accumulate toxic levels of ammonia which is then excreted into the surrounding waters possibly acting as the killing agent in blooms (Okaichi and Nishio 1976; Fukuyo et al. 1990). Extensive toxic blooms have been reported off the east and west coasts of India, where it has been implicated in the decline of fisheries (Aiyar 1936; Bhimachar and George 1950).

3.2.5 Cochlodinium polykrikoides *Margelef, 1961*

Cochlodinium polykrikoides is an unarmoured, marine, red tide-producing marine planktonic dinoflagellates species, generally found in aggregations of four or eight cell zooids (as shown in Fig. 3.2). This is a highly motile species of known for causing fish kills around the world and well known for fish kills in marine waters of Southeast Asia (Japan and Korea). The genus *Cochlodinium* was first identified in 1895 by Schütt (1895) and has been forming harmful algal blooms in the coastal waters of Southeast Asia and North America for many decades. Harmful algal blooms (HABs) caused by the toxic dinoflagellate *Cochlodinium polykrikoides* have been recorded across Asia, North America and even Europe over two decades. Production of resting cysts and subsequent transport via ships' ballast water or/and the transfer of shellfish stocks could facilitate this expansion.

The past two decades have seen *Cochlodinium* blooms expanded in their geographic distribution across Asia, Europe and North America, with fisheries losses associated with blooms in South Korea alone exceeding $100 M annually (Kim 1997). More than 40 species of *Cochlodinium* have been described, although the two primary HAB-forming species are *C. polykrikoides* and *C. fulvescens*. Both of these species are large (~40 μm) athecate dinoflagellates that commonly form chains of 2–16 cells. *Cochlodinium* blooms are generally characterized by spatially large (10–100 s of kilometres) and dense (>1000 cells ml^{-1}) cell aggregates that are heterogeneous in their vertical and horizontal distributions. These blooms are strongly ichthyotoxic and can also kill many other marine organisms, although the compound(s) responsible for these impacts have yet to be identified and bloom-associated toxins are not known to affect human health. Partly due to the recent expansion of *Cochlodinium* blooms and the general difficulty in culturing this species, there is far less known about the autecology and toxicity of *Cochlodinium* compared to other HAB species, particularly for the recently described *C. fulvescens*. With this review we seek to characterize the current state of knowledge regarding taxonomy, phylogeny, detection, distribution, ecophysiology, life history, food web interactions and mitigation of blooms formed by *Cochlodinium* as well as point out pressing questions regarding this increasingly important HAB genus (Kudela and Gobler 2012).

As with many of the unarmoured dinoflagellates, morphological features vary as a function of environmental conditions and life cycle. Typical preservation methods can distort or even destroy cells, leading to under-representation and misidentification of organisms with morphology similar to *Cochlodinium* such as *Gymnodinium catenatum* and *Gymnodinium impudicum* (e.g. Cho and Costas 2004; Curtiss et al. 2008; Howard et al. 2012). It is, perhaps, not surprising that the taxonomy of even the most widely studied *Cochlodinium* species is unclear. The genus *Cochlodinium* was established more than a century ago with the identification of *C. strangulatum* (Schütt) Schütt (1895). The majority of reported organisms within this genus (50 species, with 40 accepted; Guiry and Guiry 2011) are rare heterotrophic organisms and not well studied or described. Within this large group, only four species are known to produce chloroplasts and form chains: *C. polykrikoides* (synonymous with C. catenatum Okamura 1916 and *C. heterolobactum* Silva, 1967), *C. fulvescens, C. geminatum,* and *C. convolutum* (Iwataki et al. 2007; Matsuoka et al. 2008). Of those species, only two, *C. polykrikoides* and *C. fulvescens*, are confirmed ichthyotoxic organisms.

The holotype for *C. polykrikoides* was originally described by Margalef (1961) based on isolates from Puerto Rico, although there is some dispute about the proper naming of the species since *C. catenatum* Okamura, 1916 was described first, with the two organisms now accepted to be subclades of the same species (c.f. Matsuoka et al. 2008). The genus is almost certainly polyphyletic (Iwataki et al. 2010) with the sole diagnostic character being that the cingulum surrounding the cell more than 1.5 times (Kofoid and Swezy 1921; Iwataki et al. 2007, 2010). Besides the characteristic cingulum encircling the cell twice, *C. polykrikoides* has a reddish-orange eyespot in the anterior dorsal part (epicone) of the cell, with multiple rodlike chloroplasts. As described, the holotype is ~50 μm in length and typically forms chains of up to 16 cells in length. *C. catenatum* Okamura, 1916, was described as a smaller (21–26 mm in length), chain-forming dinoflagellate with the cingulum encircling the cell more than 1.5 times, with a smaller eyespot located in the epicone and with multiple rod-like chloroplasts. *C. catenatum* Kofoid and Swezy (1921) is superficially similar to *C. catenatum* sensu Okamura but was described as heterotrophic with a more central nucleus. A third organism, *C. convolutum* Kofoid and Swezy, 1921, is larger (60–70 mm in length), lacks an eyespot, has reticulate chloroplasts together with many scattered small grains at the periphery of the cell, and rarely forms chains. *C. convolutum* has been reported to form red tides (discolourations of the surface water) in both Japan and California but has not been implicated in fish or shellfish mortality (c.f. Matsuoka et al. 2008). The most recently described species in this genus, *C. fulvescens* Iwataki et al. (2007), is ~40–50 μm in length, forms chains of 2–4 cells (up to 8 cells have been observed; Kudela, pers. obs.), has a cingulum that wraps 2X around the cell (similar to *C. polykrikoides*), has a reddish eyespot in the dorsal epicone, and exhibits granular chloroplasts. This organism has been difficult to culture, and there is relatively little known about its ecophysiology or the mode of action for harmful effects, but blooms in British Columbia, Canada and Central California have been linked to mortality of aquacultured salmon and abalone (Whyte et al. 2001).

In late 2008 and early 2009 (November–February), there was a massive bloom of *Cochlodinium polykrikoides* in the Sea of Oman, off the coast of Oman in the Arabian Sea observed by Matsuoka et al. (2010) and Richlen et al. (2010). It was notable for being based on *Cochlodinium polykrikoides* rather than the *Noctiluca scintillans* (*Noctiluca miliaris*) that had been more usual in the immediately previous years (Parab et al. 2006; Al-Azri et al. 2014). The bloom resulted in massive dying off of fish, damage to coral reefs and interference with desalinization plants (Al-Azri et al. 2014).

Red tides dominated by *Cochlodinium polykrikoides* often lead to great economic losses, and some methods of controlling these red tides have been developed. Lim et al. (2017) devised an ecofriendly measure to control of ichthyotoxic *Cochlodinium polykrikoides* using the mixotrophic non-toxic dinoflagellate *Alexandrium pohangense*, which grows well mixotrophically feeding on *C. polykrikoides*. Most significantly, the populations are maintained by photosynthesis, and incubation of *A. pohangense* did not result in fish mortalities. Hence, compared to other methods, this controlling measure seems to be much safer and sustainable one which can be effectively applied for aquaculture.

Tang and Gobler (2012) made visual confirmation of the production of resting cysts by *C. polykrikoides* in laboratory cultures which subsequently germinated up to 1 month. This evidence of resting cyst production by the species provides a mechanism to account for the recurrence of annual blooms in given locales and global expansion of blooms made by the species during the past two decades.

HABs are recurring phenomenon in the coastal waters of Kota Kinabalu, Malaysia, caused by *Pyrodinium bahamense* var. *compressum*, *Gymnodinium catenatum* and *Cochlodinium polykrikoides* (Adam et al. 2011). The authors did extensive samplings covering southwest monsoon (SWM), northeast monsoon (NEM) and Inter-Monsoon (IM) periods, and it is revealed that rainfall is the prime factor for triggering bloom formation by carrying huge nutrients.

In addition to rainfall, sea surface temperature (SST) and photosynthetically active radiation (PAR) have important roles in algal bloom formation in aquatic systems. PAR refers the spectral range (wave band) of solar radiation (400–700 nm) used by phytoplankters for photosynthesis. SST has been shown to be a potential factor in the growth of marine dinoflagellate *Cochlodinium polykrikoides* and thus determining when blooms form (Kim et al. 2016). Laboratory studies have shown that *C. polykrikoides* have the most significant growth between 25.0 and 26.0 °C (Lee and Choi 2009; Griffith and Gobler 2016). There should be enough light for this phytoplankton to photosynthesize as required for other phytoplankters. Studies have proven that *C. polykrikoides* have higher growth rates when solar insolation is increased (Tomas and Smayda 2008). In addition, currents also play an important role when transporting toxic microalgae to favourable areas for a bloom to spawn (Kim et al. 2016).

3.2.6 **Karenia mikimotoi** ***Miyake & Kominami ex Oda Gert Hansen and Ø.Moestrup, 2000***

The unarmoured dinoflagellate *Karenia mikimotoi* (Miyake & Kominami ex Oda; Gert Hansen & Ø.Moestrup) (also known in the literature as *Gyrodinium aureolum*, *Gymnodinium mikimotoi* and *G. nagasakiense*) has a widespread global distribution (Brand et al. 2012). Mass mortality of finfish, shellfish and other benthic organisms was encountered due to bloom of this species throughout the world, including Ireland, Scotland, Norway, India, Korea and Japan (see Davidson et al. 2009; Okaichi 1989; Ottway et al. 1979; Park et al. 1989; Robin et al. 2013; Tangen 1977). The species is also found within temperate coastal waters of the Pacific (Faust and Gulledge 2002), the North West Atlantic (Mahoney et al. 1990) and the North East Atlantic off the coasts of Norway (Dahl and Tangen 1993) and the Irish Sea (Evans 1975). It is particularly common in the western English Channel where cells usually peak in abundance between June and September (Pingree et al. 1975; Holligan et al. 1984; Garcia and Purdie 1994). Extensive blooms have been reported in offshore United Kingdom (UK) waters in recent years including the western English Channel in 2003 (Vanhoutte-Brunier et al. 2008), the west coast of Ireland in 2005 (Silke et al. 2005) and the northern Scottish waters in 2006 (Davidson et al. 2009). *K. mikimotoi* blooms can reach high densities of >103 cells mL^{-1} (>10 mg m^{-3} Chl a) with dramatic effects for the marine ecosystem (Gentien 1998).

At local to regional scales, coincidence of large-scale mortalities for fish and other aquatic organisms has been related to *Karenia* spp.; nevertheless, it is not quite evident to what extent the biota and habitats are affected by the toxicity related to large blooms (Silke et al. 2005). The observed mortalities might be attributed to both ichthyotoxic compounds (Satake et al. 2005) and deoxygenation (Tangen 1977) although it is quite uncertain which of these is the main harmful agent. It is no surprise that extensive research is directed mainly towards the detection (Shutler et al. 2012; Yuan et al. 2012) and prediction of *Karenia* spp. blooms to a lesser extent. Importantly, such blooms carry significant effects over larger scales, most notably their (temporal/spatial) contribution to ecosystem productivity. The substantial increase in biomass and primary productivity can heavily dominate the overall season, from occurrence of *Karenia brevis* blooms in West Florida waters (Bendis et al. 2004; Hitchcock et al. 2010) for a short duration. Recently Soto et al. (2018) recorded HABs of *K. brevis* along gulf coastal regions and observed recurrence of intense large blooms (in Florida and Texas) whereas very rare occurrence in Mississippi coastal waters. They had observed westward advection of *K. brevis* blooms.

K. brevis is a dinoflagellate which contains the photosynthetically active pigment chlorophyll a (Chl *a*) and is restricted to the Gulf of Mexico and the Caribbean but has been carried by ocean currents around Florida and up the east coast of the United States as far as North Carolina. *K. brevis* usually blooms in the late summer and autumn, almost every year off the west coast of Florida, causing mass killing of fish and bird. During a red tide event or *Karenia brevis* bloom, the water is discoloured to a reddish-brown hue. Associated with these algal bloom episodes of *K. brevis*, a variety of phytoplankton-related natural toxins have been identified. The most important group is the neurotoxic brevetoxins (i.e. *Ptychodiscus brevis* toxin, PbTx). As a group, the brevetoxins are lipid-soluble, cyclic polyethers. Over nine different brevetoxins have been isolated in seawater blooms and *K. brevis* cultures, as well as multiple analogues and derivatives from the metabolism of shellfish and other organisms. This is important because the potent neurotoxins can be transferred through the food web where they affect and even kill the higher forms of life such as zooplankton, shellfish, fish, birds, marine mammals, and even humans that feed either directly or indirectly on them due to bioaccumulation.

A combined set of factors like cellular toxicity, production of mucilage that can cause respiratory difficulties and oxygen depletion are related for mortalities, albeit the relative contribution of individual factor is still known. *Karenia mikimotoi* is known to produce a set of toxins with cytotoxic properties such as hemolysin (Neely and Campbell 2006), gymnocins A and B (Satake et al. 2002, 2005) and polyunsaturated fatty acid (PUFA) (Mooney et al. 2007; Parrish 1987). However, the true mechanism of toxicity is unknown.

Karenia mikimotoi is known to produce toxins with cytotoxic properties including hemolysin (Neely and Campbell 2006), gymnocins A and B (Satake et al. 2002, 2005) and polyunsaturated fatty acid (PUFA) (Mooney et al. 2007; Parrish 1987). Respiration by *Karenia mikimotoi* cells and bacterial respiration associated with the breakdown of the bloom and decaying macro-organisms increases biochemical oxygen demand (BOD). Deficiency in oxygen in bottom waters in estuaries and coastal bays is observed. In extreme cases, either hypoxic (levels of dissolved oxygen near or below 2.0 mg l^{-1} O_2) or anoxic (total depletion of oxygen), leading mass mortalities of benthos are observed (Diaz and Rosenberg 2008). The process is exacerbated if the water column is stratified, either by freshwater inputs to estuaries or through seasonal heating of surface waters. *Karenia mikimotoi* cells can secrete mucous made of an extracellular polysaccharide which may act as an additional factor in causing mortality. It is thought that cells secrete this substance to reduce shear which reduces the probability of individual cells colliding with one another, which in turn reduces the incidence of autotoxicity (Gentien et al. 2007).

The species is thought to overwinter in low numbers as motile cells awaiting favourable bloom conditions (Gentien 1998). Blooms are highly variable in timing, duration, spatial extent and magnitude and do not occur every year (García-Soto and Pingree 2009); thus the precursor and optimum blooming conditions of *Karenia* spp. are poorly understood. *K. mikimotoi* is a slow-growing, shade-adapted species that is occasionally able to compete out much faster-growing diatoms (Gentien 1998). While prevailing hydrography (e.g. temperature or salinity) may regulate the regional distribution of a particular species and contribute to the timing of the annual/seasonal maxima, it is generally thought that light or nutrient availability determines the growth, biomass and duration of local scale blooms. Precursor causes of large *K. mikimotoi* blooms are suggested to include enhanced growth by sunlight-driven phototaxis, rainfall- or benthic-mediated nutrient availability and concentration processes such as wind-driven advection and frontal systems (Gentien 1998). However, these remain to be substantiated or proven.

K. mikimotoi neither co-occurs with (positive relationship) nor inhibits (negative relationship) other species. Our observations suggest that *K. mikimotoi* may have an important ecological function in the western English Channel acting as an important intermediary in the carbon cycle. *K. mikimotoi* can exhibit subsurface blooms near and at the thermocline (Holligan et al. 1984) in the latter stages of the bloom and can also migrate through the water column (Barnes et al. 2015).

For the formation of *K. mikimotoi* bloom, the precursor conditions are mainly concerned to increase in rainfall and subsequent decrease in salinity together with enrichment of nutrient concentrations (as traced by silicate). However, riverine outflow could provide sufficient nutrients in the surface layer for *K. mikimotoi* growth through two devices: (1) directly through nutrient-rich river water and (2) indirectly through changes in the buoyancy structure leading to increased nutrient diffusion across the nutricline. As *K. mikimotoi* is a motile species, increasing the nutrient diffusion may give it an advantage, although our data (surface nutrients only) cannot distinguish between these two processes (Barnes et al. 2015).

3.2.7 **Coscinodiscus** *sp. Ehrenberg, 1839*

The *Coscinodiscus* sp. was established by Ehrenberg in 1839 based on the morphology and size of frustule (Sar et al. 2010). The complex frustule was composed of two thecae: an epitheca and a hypotheca. They have delicate nanostructure, comprising of valve and several copulae (cingulum), and the cingula were formed. The valves of *Coscinodiscus* sp. have irregular round plates and non-uniform sizes, which are difficult to arrange using existing self-assembly methods (Wang et al. 2012). Cells are disc-shaped, cylindrical or wedge-shaped, and solitary and cell size: diameter (apical axis) ranges from 30 to 500 μm. This is considered as one of the largest marine planktonic centric diatom genera (400–500 taxa described and identified) so far and widely distributed in warm water to boreal. Although primarily a coastal species, it is also found beyond the continental shelf and in estuaries too.

3.2.8 Trichodesmium erythraeum *Ehrenberg and Gomont, 1892*

Trichodesmium (sea sawdust), a genus of blue-green algae taxonomically assigned to the order Oscillatoriales, is a planktonic, marine diazotrophic, filamentous cyanobacteria characterized by trichomes. They normally occur in macroscopic bundles (colonies) and widely distributed across oligotrophic (nutrient-poor) pelagic waters (Capone et al. 1997; Carpenter et al. 2004). Representatives within the genus have consistently been shown to be stable components in tropical and subtropical waters, particularly in the Atlantic, eastern tropical Pacific and Indian Oceans. They form massive, dense and intensive surface blooms in favourable environmental conditions, visible to the naked eye, and it is reported that these contribute >30% of algal blooms of the world. The water becomes red or brownish-yellow because of the presence of cyanobacteria-specific phytopigments, such as phycoerythrin and photoprotective carotenoids. These blooms are extremely patchy in nature, mainly concerned to the physical variability of the water bodies. The blooms function as a habitat for an array of heterozygous organisms, such as other cyanobacteria, bacteria, eukaryotic microalgae, protozoa, fungi and copepods (a crustacean arthropod).

This diazotrophic cyanobacterial genus *Trichodesmium* has tremendous contribution to the nitrogen influx of the global marine ecosystem and has paramount ecological importance in the context of ocean nitrogen budget. The species has substantial contribution for sustenance of marine community through active release of key nutrients (such as carbon and nitrogen) and upon death and decay; hence this photoautotrophic genus plays a crucial role in the biogeochemical cycling of basic elements in marine ecosystems. *Trichodesmium* forms extensive surface blooms under favourable conditions that are easily visible due to their dense accumulation at the surface. *Trichodesmium erythraeum* forms the most common bloom-forming species (Sellner 1997; Capone et al. 1998) among five identified species. Trichodesmium has been intensively studied due to its significant role in biogeochemical cycling and ecological importance, but the toxic nature of this genus remains scarcely understood (Kerbrat et al. 2010, 2011). Sellner (1997) and O'Neil et al. (2012) have reviewed *Trichodesmium*-induced toxicity and ecological damages caused by these blooms across the globe, and type of toxins has also been documented. *Trichodesmium* blooms have been reported to cause only.

Mass-scale fish mortality has been recorded in Indian waters (D'Silva et al. 2012), but none of these blooms were investigated for the toxins released by them. Thus, there is an urgent need to study the type of toxicity caused by these blooms and the toxins involved by using simple and sensitive techniques. A single assay or test is not good enough to monitor or study the cyanobacteria toxicity to assess the possible health risk for humans (Hisem et al. 2011). Hence, multiple assays would be the best option which can cover a broad array of study, including invertebrates and mammalian cell model (Narayana et al. 2014).

3.3 Category of Harmful Algal Blooms (HABs)

Human poisoning shellfish-borne toxins can be categorized into the following six major pathologies all derived from HABs of diverse ecotoxicological relevance, which may pose the great continuing threat to human health. Each category of HABs is characterized by specific potent toxin of variable toxicity and clinical symptoms that are expressed after people consumed contaminated finfish and shellfish.

3.3.1 Paralytic Shellfish Poisoning (PSP)

Paralytic shellfish poisoning (PSP) is a potentially lethal clinical syndrome causing important public health risk from dinoflagellate-bivalve unique coupling. It is caused by eating bivalve molluscs (mussels, scallops and clams) contaminated with a group of structurally related marine toxins, best-known paralytic shellfish toxin (PST). Structurally the toxin of this class could be divided into two major groups: saxitoxin (STX; $C_{10}H_{19}N_7O_4$) (Shumway 1990; James et al. 2010) and neosaxitoxin (NSTX); both are potent natural neurotoxic alkaloids. STXs are accumulated in siphons and hepatopancreas of the filter-feeding (straining particulates suspended in water) mollusc shellfishes by ingesting dinoflagellates. Thus they act as potential vectors for transfer of STXs and other toxic materials from microalgae to higher trophic levels (e.g. birds, marine mammals and human), causing severe ecological and economic disasters. Interestingly, STXs are non-toxic to the shellfish (including the commercially exploited species) and may be retained for a long time without harmful effects, because they use non-acid digestion. But when we eat these shellfish, the acid media in human stomach break down some of the long molecules produced by the dinoflagellates into shorter toxic ones. These toxic algae are also ingested by selective zooplanktons besides the bivalve molluscs as previously discussed. It is worth to mention that cooking of the contaminated shellfish in higher temperature does not destroy the toxin.

The dinoflagellate *Alexandrium* is one of the most widespread HAB genera, and its success is based on key functional traits like allelopathy, mixotrophy, cyst formation and nutrient retrieval migrations. Recently, Natsuike et al. (2017) introduced a simple mesocosm method to investigate the seasonal changes in germination fluxes of the resting cysts, specifically *Alexandrium fundyense* (*A. tamarense* species com-

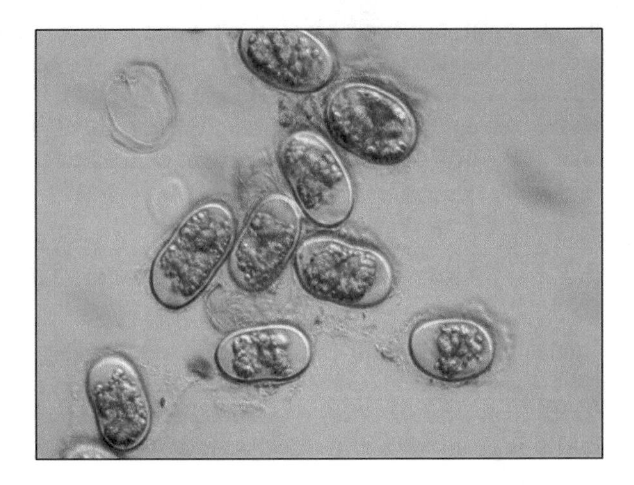

Fig. 3.3 Cysts of *Alexandrium tamarense*. (Courtesy: Dr. Mindy L. Richlen, National Office for HABs, WHOI, USA)

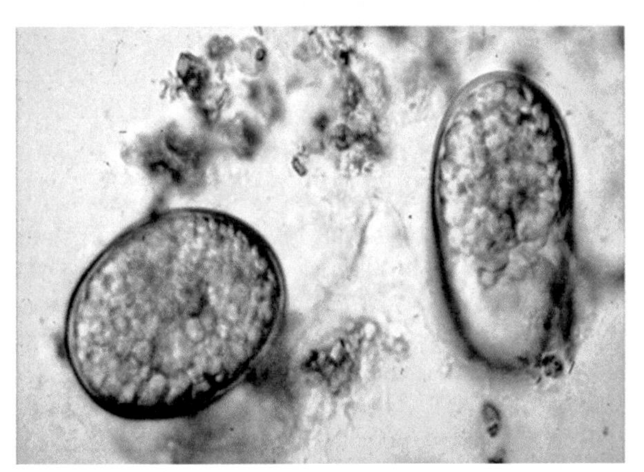

Fig. 3.4 Cysts of *Alexandrium fundyense*. (Courtesy: Dr. Mindy L. Richlen, National Office for HABs, WHOI, USA)

plex Group I) and *A. pacificum* (*A. tamarense* species complex Group IV), in Kesennuma Bay, Japan (Fig. 3.3). The study revealed the seasonal dynamics of the two species cyst germination and their bloom occurrences in the water column. Blooms occurred 1–2 months after peak germination, which strongly suggests that initial population formation by cyst germination and its continuous growth in the water column most likely contributed to toxic bloom occurrences of these species (Katsuhide et al. 2017).

Different life stages of two mating-compatible clones of the paralytic shellfish toxin (PST)-producing dinoflagellate *Alexandrium fundyense* Balech were separated by Persson et al. (2012) adopting a combination of techniques; toxin profile changes were shown to occur very quickly in both the strains GTX1 and GTX2 during gamete formation (Fig. 3.4). Loss of toxins to the environment would lead to free toxins surrounding the bloom (Persson et al. 2012).

Recurrent blooms in coastal regions of the central and northern Baltic Sea have been recorded by Hakanen et al. (2012) mainly made by the potentially toxic dinoflagellate *Alexandrium ostenfeldii* (Figs. 3.5 and 3.6). They monitored the bloom and

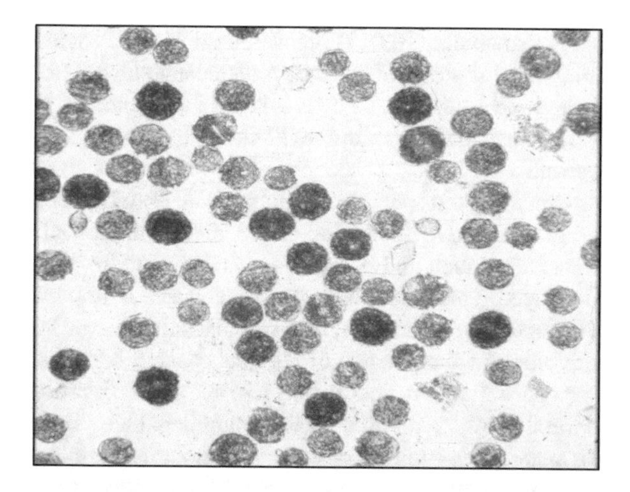

Fig. 3.5 High concentration of *Alexandrium* cell. (Courtesy: Dr. Mindy L. Richlen, National Office for HABs, WHOI, USA)

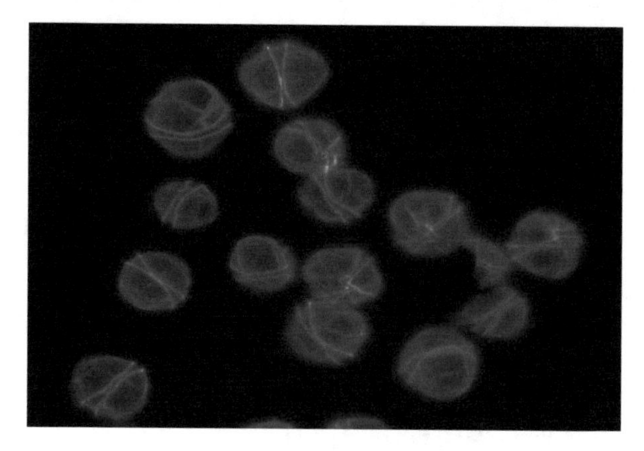

Fig. 3.6 *Alexandrium* cells stained with Calcofluor. (Courtesy: Dr. Mindy L. Richlen, National Office for HABs, WHOI, USA)

toxin dynamics which was the first evidence of PSP toxins produced by the dinoflagellate species in this region in warm water conditions (around 20 °C). Abundance of *the species* was significantly related to high concentrations of resting cysts in the sediment, causing localized blooms in shallow areas (Hakanen et al. 2012).

Since 2012, dense *Alexandrium ostenfeldii* blooms have recurred annually in a creek located in the southwest of the Netherlands, an area characterized by intense agriculture and aquaculture. Over 3 consecutive years, Brandenburg et al. (2017) investigated how the physico-chemical and biological factors influenced *Alexandrium ostenfeldii* bloom dynamics. Low salinities accompanied by increased wind speed corresponded to a delayed *A. ostenfeldii* bloom with reduced population densities. Highest population densities generally corresponded to high temperatures, low DIN:DIP ratios and low grazer densities while comparing year wise data. The authors have recorded the key role of nutrient availability, absence of grazing and particularly the physical environment on the magnitude and duration of *A. ostenfeldii* blooms (Brandenburg et al. 2017).

Borkman et al. (2012) recovered an unusual toxic *Alexandrium peruvianum* by studying its morphology and molecular analysis from Narragansett Bay, Rhode Island, USA, and detected six saxitoxin congeners (GTX2, GTX3, B1, STX, C1 and C2) along with gymnodimine (12-methyl gymnodimine) and spirolide (13-desmethyl spirolide C) toxins. Discovery of this species is significant as it appears to be an emergent bloom species in a global scale (Borkman et al. 2012).

In a bioactive metabolite of the guanidine derivative, the toxin is stored within the bivalve shell and not released into the environment until the cell is crushed or destroyed. This potent neurotoxin interferes with nerve impulse transmissions (i.e. conductance of nerve signals) in poikilothermic or homoeothermic vertebrates by inhibition of the passage of Na ions causing neuromuscular paralysis. PSP toxins are concentrated in shellfish as a result of the filtration of toxic algae produced by several key 'red tide'-related dinoflagellate species such as *Alexandrium*, *Gymnodinium* and *Pyrodinium*. Most notably, members of these three genera cause PSP by producing a suite of neurotoxins called paralytic shellfish toxins (PSTs). These toxins interfere with the membrane-associated sodium channels of excitable cells (such as muscle and neurons) and inhibit nervous transmission, causing temporary neuromuscular paralysis (Narahashi and Moore 1968). Predators of bivalve shellfish (scavenging shellfish, lobsters, crabs and fish) may also be vectors for saxitoxins, thus expanding the potential for human exposure (Halstead and Schantz 1984). Geographically, the most risky regions for PSP are temperate and tropical marine coasts. In a genus *Alexandrium* causing serious episodes of PSP in coastal waters of South America, PSP has been recognized as an important public health risk for over a century (Hallegraeff 1993; Balech 1995; Lagos 2003). In North America, this includes Alaska, the Pacific Northwest and the St. Lawrence region of Canada; however, incidents of PSP regularly occur in the Philippines and other tropical regions. Toxic shellfish have also been found in temperate regions of southern Chile, England, Japan and the North Sea.

Alexandrium tamarense is a toxic marine dinoflagellate causing PSP. Because of the morphologically resembling species with *A. tamarense*, *A. catenella* and *A. fundyense*, they are grouped under *A. tamarense* species complex (Anderson et al. 2012) which were previously confined in Europe, North America and Japan during the 1970s (Dale and Yentsch 1978). Their geographical distribution accessed to a wide coverage from the subtropical to the subarctic of the north hemisphere and into the temperate south hemisphere (Hallegraeff 1993; Lilly et al. 2007) which may be due to impact of climate change (Hallegraeff 2010). Frequency of *A. catenella* bloom was associated with low temperatures, but not with salinity, Chl a concentration and presence of predators (measured as clam biomass) in southern Chile (Chilean Inland Sea) (Diaz et al. 2014). It is estimated that there are 1600 annual cases of PSP worldwide and an estimated 300 fatalities among these cases.

The initial symptoms of PSP are numbness or tingling on the lips, tongue and face within 10 min to 2 h after shellfish consumption with red tide organisms. The timing of symptom onset is thought to be dose dependent (Gessner et al. 1997a; McLaughlin et al. 2011). Depending on the amount of toxin consumed, symptoms may aggravate resulting in tingling of fingers and toes and loss of control of arms

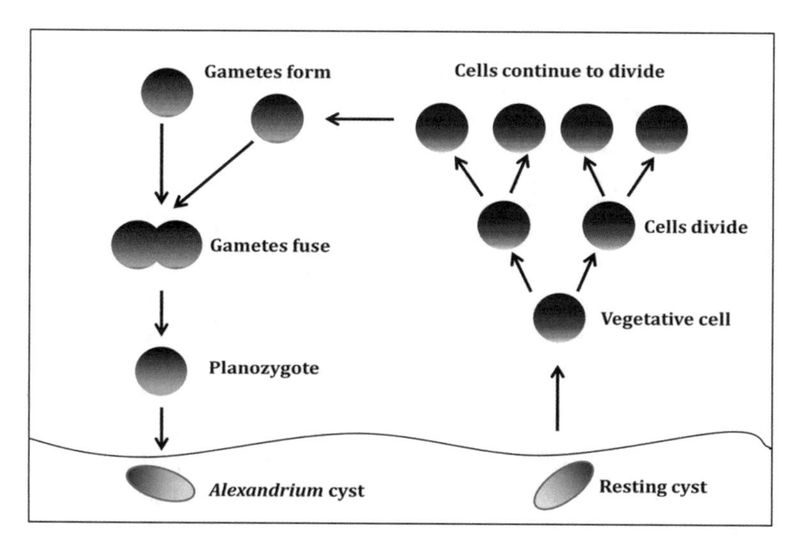

Fig. 3.7 Schematic diagram of the *Alexandrium* sp. Life cycle

and legs, followed by severe respiratory problems in a person who is not suffering from any breathing trouble (Gessner et al. 1997b), and even death also occurs.

In severe cases, other gastrointestinal symptoms are vomiting, abdominal pain, water diarrhoea, nausea and hypersalivation along with a wide range of neurologic symptoms. In addition, the wide range of neurologic symptoms may include weakness, numbness, dizziness, dysarthria (motor speech disorder), paresthesia (abnormal tingling sensation), double vision, loss of coordination, vertigo or dizziness and/or a 'floating' sensation. Recently, Knaack et al. (2016) suggested that dysphagia (swallowing disorder) and dysarthria (a speech disorder in which the pronunciation is not clear) are most likely the strongest indicators of PSP. Generally, recovery rate is rapid and complete with most symptoms resolving within 24–72 h, with 14 days representing the maximum recovery window (Rodrigue et al. 1990; Gessner et al. 1997a).

3.3.1.1 General Features of Life Cycle of *Alexandrium* sp.

The sexual life cycle of *Alexandrium* species, as depicted in Fig. 3.7, includes a dormant benthic stage (resting cyst). A role for benthic cysts has been cited in species dispersion, resistance to unfavourable conditions, population resilience through sexual recombination and bloom onset and termination (Anderson et al. 1982; Anderson 1984; Garcés et al. 1999; McGillicuddy et al. 2003). Approximately 13–16% of living dinoflagellates produce resting cysts as a part of their life history (Head 1996), and knowledge of their cyst stages allows better understanding of the mechanism of occurrence, persistence and disappearance of many red tide blooms. Moreover, the toxicity of resting cysts can be higher than that of vegetative cells; resting cyst formation must be taken into account when analysing the possible effects of a bloom on human health (Dale et al. 1978; Lirdwitayaprasit et al. 1990; Oshima et al. 1992). The roles played by resting cysts in the life cycle and ecology of dinoflagellates species can be very different depending on the species. For example, the life cycles and encystment strategies of *Alexandrium* species are both complex and species-specific (Figueroa et al. 2006, 2008). Environmental factors were responsible affecting the rate of excystment and maturation in *Alexandrium minutum* and *A. tamarense*, as experimentally confirmed by Rathaille and Raine (2011).

Dinoflagellate cysts (dinocysts) form because of sexual fusion of gametes, which are formed in response to specific conditions, such as low temperature, disturbance, metal contamination and nutrient depletion (Head 1996; Tang and Gobler 2012) along with the coarse sediments, high sedimentation rates and high anthropogenic disturbances for low cyst production (Lu et al. 2017). The life cycle begins when a resting cyst germinates to form a vegetative cell; these cells divide multiples rapidly through asexual means (without sex cells) during its productive stage which results in bloom. Eventually these reproduce sexually by forming gametes, which fuse to form a planozygote. The planozygote resembles the vegetative cell except for possessing two longitudinal flagella; it subsequently transforms into a cyst, completing the life cycle. The cysts, as resting or dormant stage, 'hibernate' in the sediment until activated by favourable conditions in the next outbreak. Dinocysts are formed of very resistant organic materials and are generally well preserved in sediments.

Exogenous (environmental) factors mainly affect the timing and success of germination, whereas endogenous (physiological) factors regulate the germination of resting cysts and determine whether this stage is involved in the short- or long-term survival of the species or serves as a bloom maintenance mechanism. Local hydrographic and environmental factors greatly affect the strength and timing of sexual induction, and therefore the success of encystment and germination, while hydrodynamic processes determine the location of cyst, deposits (Anderson et al. 2005; Keafer et al. 2005; McGillicuddy et al. 2005; He et al. 2008). Within the genus *Alexandrium*, the dormancy requirements of *Alexandrium tamarense* cysts differ between deep water and shallow coastal environments (Anderson and Keafer 1987; Matrai et al. 2005). In *Alexandrium catenella*, dormancy periods are highly variable, ranging from 28 to 97 days (Yoshimatsu 1984; Hallegraeff et al. 1998). In Mediterranean populations of this species, they are characterized by a gradual rather than a synchronous pattern of germination (Figueroa et al. 2006) that allows for rapid cycling between benthic and planktonic stages (Hallegraeff et al. 1998). Recently, Lu et al. (2017), in the context of impacts of metal contamination, demonstrated that both *Alexandrium* and *Diplopsalis* cysts were sensitive to metal contamination; however, *Gyrodinium, Pheopolykrikos* and *Lingulodinium* cysts had high resistance to metal contamination.

Fig. 3.8 A *Dinophysis* bloom in Norway coastal waters. (Courtesy: Dr. Mindy L. Richlen, National Office for HABs, WHOI, USA)

3.3.2 Diarrheic Shellfish Poisoning (DSP)

Diarrheic shellfish poisoning is characterized by acute gastrointestinal symptoms triggered by the ingestion of shellfish contaminated with okadaic acid and related toxins. A group of bivalve molluscs (such as mussels, clams, scallops and oysters) are the most common vectors for the DSP toxins which are produced by a community of dinoflagellates, most notably *Dinophysis* spp. and *Prorocentrum* spp. (James et al. 2010; Valdiglesias et al. 2011). Outbreaks of DSP have been reported from all parts of the world such as the United States, Japan, France and other parts of Europe, Canada, New Zealand, the United Kingdom and South America (Yasumoto et al. 1978; Kawabata 1989; Belin 1991; van Egmond et al. 1993; Hinder et al. 2011; Trainer et al. 2013). A dinophysis bloom in Norway coastal water has been depicted in Fig. 3.8.

Large number of biotoxin-producing dinoflagellates *Dinophysis* sp. leading to outbreak of DSP was recorded by Whyte et al. (2014) around the Shetland Island, Scotland, in 2006 and 2013, coinciding with a change in the prevalent wind direction and speed (Fig. 3.9). Wind direction and speed in the North East Atlantic and the North Sea is strongly influenced by the North Atlantic oscillation (NAO) with a positive relationship between it and wind direction. It has been noted that a positive trend in the NAO is linked to climate change. Analysis of wind patterns therefore acts as a potential measure for early warning of future bio-toxicity events. DSP occurs even at low concentration ($< 10^3$ cells l^{-1}) due to presence of diarrhetic toxins

Fig. 3.9 SEM image of *Dinophysis acuminata*. (Courtesy: Dr. Mindy L. Richlen, National Office for HABs, WHOI, USA)

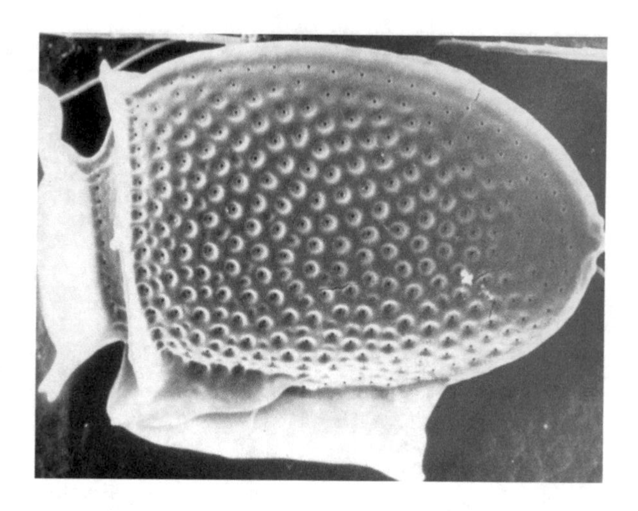

(okadaic acid and dinophysis toxin). Toxicity and toxin profiles are variable more between strains than species. Shellfish contamination results from a complex balance between fish selection, adsorption, species-specific enzymatic transformation and allometric processes.

Okadaic acid (OA) and its analogues, the dinophysistoxins (DTXs), are lipophilic marine toxins produced by several phytoplanktonic species causing DSP in humans. DSP toxins are considered important phytotoxins since their presence produce health effects in human consumers along with severe economic losses. It is produced principally by a group of dinoflagellates of the genus *Prorocentrum* (mainly *P. Lima* followed by *P. concavum*) and *Dinophysis* (mainly *D. acuta*, *D. acuminate* and *D. fortii*) (FAO 2004). OA, like other phycotoxins, is accumulated by shellfish, mainly bivalve molluscs – such as mussels, scallops, oysters or clams – and several fishes through consuming microalgae. This is subsequently consumed by humans causing alimentary intoxications. The characteristics of these toxins are not changed by cooking or freezing, and they do not modify the taste of the contaminated organisms, hence very difficult to detect (McCarron et al. 2008; Reboreda et al. 2010).

Clinical symptoms: DSP diagnosis is largely made by dietary history and symptoms, similar to other shellfish illness. Symptom onset typically occurs within 30 min to 4 h after eating contaminated shellfish. The main symptom is incapacitating diarrhoea, followed by nausea, lethargy, general weakness, vomiting and abdominal cramps (James et al. 2010). The symptoms may be severe and lead to dehydration but are usually self-limiting and continue for about 3 days. DSP outbreaks reported in temperate seas in particular, in Chile, Europe and Japan. The first record of DSP was recorded in 1997 when approx. 50 people became ill after consuming shellfish (Hinder et al. 2011). DSP is widespread and has been reported from 11 countries across Europe (Denmark, France, Germany, Ireland, Italy, Norway, Portugal, Spain, Sweden and Netherland) (Smayda 2006). The severe outbreak occurred in Spain during 1981 when 5000 people were affected mostly from Madrid (Fraga et al. 1984, 1988; Gestal-Otero 2000).

Fig. 3.10 Sea lion affected by domoic acid. (Courtesy: Dr. Mindy L. Richlen, National Office for HABs, WHOI, USA)

3.3.3 Amnesic Shellfish Poisoning (ASP)

Domoic acid (DA; $C_{15}H_{21}NO_6$) is a naturally occurring excitatory potent neurotoxic amino acid produced by pennate marine diatom genera *Pseudo-nitzschia* and *Nitzschia* causing domoic acid toxicity (DAT). The potential risk of domoic acid poisoning (DAP) to human health was discovered in 1987 in eastern Canada (Perl et al. 1990a, b; Teitelbaum 1990; Teitelbaum et al. 1990). Distributed from tropical to polar regions, over 35 species of *Pseudo-nitzschia* have been identified, some of which produce domoic acid (DA). This neurotoxin is responsible for the intoxication syndrome in humans and marine animals called amnesic shellfish poisoning (ASP), in shellfish, sardines and anchovies. Most of these toxigenic species are an important component of phytoplankton and have been recorded in the Southern Cone of South America (Argentina, Brasil, Chile and Uruguay).

DA toxin is produced by blooms of *Pseudo-nitzschia* sp. Shellfish and other marine organisms feed on potentially toxigenic *Pseudo-nitzschia* sp. and concentrate the toxin within them. Hence, the shellfish become harmful to wildlife (sea lion) and humans that consume them (Fig. 3.10). Widespread occurrence of DA in eight macrobenthic species has been recorded of multiple feeding strategies (i.e. filter and deposit feeders, predators, scavengers), frequently at elevated concentrations (>500 pm where the regulatory limit for human consumption is 20 ppm) (Kvitek et al. 2008). This signifies the unique pelagic-benthic coupling for toxin transfer through benthic food webs in diverse pathways. Although DA has been found in the viscera of Dungeness crab and other organisms, razor clams are one of the most significant vectors, as they can hold the toxin for up to 1 in the natural environment or several years after being processed, canned or frozen (Wekell et al. 1994). Preliminary findings from Grattan et al. (2016) raise the possibility that milder memory problems may be associated with lower level, chronic exposures in adults who are heavy consumers of razor clams. Thus, domoic acid neurotoxicity potentially may be associated with a non-amnesic syndrome.

An abrupt shift towards increasing in greater intensity and frequency of domoic acid (DA) – producing *Pseudo-nitzschia* blooms – was recorded by Sekula-Wood et al. (2011) off the California coast, Santa Barbara Basin (SBB), USA, at 540 m. This specific phenomenon might be correlated to natural shifts in climate variability coupled with a change in phase of the North Pacific Gyre Oscillation (NPGO) and its profound impact on water quality upwelled into the SBB. The NPGO emerges as the 2nd dominant mode of sea surface height variability in the Northeast Pacific, significantly correlated with fluctuations of salinity, nutrients *and* chlorophyll *a* measured in long-term observations in the California Current and Gulf of Alaska.

Along the Pacific coast of North America, from Alaska to Mexico, harmful algal blooms (HABs) have caused losses to natural resources and coastal economies and have resulted in human sicknesses and deaths for decades. Recent reports indicate a possible increase in their prevalence and impacts of these events on living resources over the last 10–15 years. Two types of HABs pose the most significant threat to coastal ecosystems in this 'west coast' region: dinoflagellates of the genera *Alexandrium*, *Gymnodinium* and *Pyrodinium* that cause paralytic shellfish poisoning (PSP) and diatoms of the genus *Pseudo-nitzschia* that produce domoic acid (DA), the cause of amnesic shellfish poisoning (ASP) in humans (Lewitus et al. 2012). These species extend throughout the region, while problems from other HABs (e.g. fish kills linked to raphidophytes or *Cochlodinium*, macroalgal blooms related to invasive species, sea bird deaths caused by surfactant-like proteins produced by *Akashiwo sanguinea*, hepatotoxins from *Microcystis*, diarrhetic shellfish poisoning from *Dinophysis* and dinoflagellate-produced yessotoxins) are less prevalent but potentially expanding.

As evidenced from the megadata generated by NOAA phytoplankton monitoring network (2001–2010), the spatiotemporal trends of *Pseudo-nitzschia* spp. were stretched from North Carolina through northern Florida along the southeastern US coastline. The species was more common in North and South Carolina while present from North Carolina to Florida. Across the majority of the Atlantic southeast US, the maximum rates of occurrence of *Pseudo-nitzschia* were recorded during late summer, early fall, with most areas experiencing the lowest rate of occurrence in the spring as observed by Shuler et al. (2012). The Outer Banks of North Carolina, however, experienced a peak of occurrence in late winter to early spring in addition to a late summer, early fall peak. *Pseudo-nitzschia* was eurythermal as it was found in temperatures ranging from <5 to 35 °C and euryhaline, tolerating a wide range of salinities (5–37). Six unique bloom events were documented during this period of 9 years, three of which contained detectable levels of domoic acid. The majority of

Fig. 3.11 A pelican killed by domoic acid. (Courtesy: Dr. Mindy L. Richlen, National Office for HABs, WHOI, USA)

these bloom events and all of the toxic events occurred in the Outer Banks of North Carolina. Given the extent and intensity of coverage afforded by the NOAA PMN, this programme provides the optimal approach to not only assess past trends but to monitor environmental changes and emerging trends in the dynamics of this toxigenic species. Understanding the dynamics of this species, resource managers would be benefitted to better predict the negative impact associated with domoic acid.

People suffered serious gastrointestinal distress and even death in extreme cases due to consumption of blue mussels harvested from the Prince Edward Island region. Some survivors were left with a permanent and profound memory disorder, called amnesic shellfish poisoning (ASP). In 1991, pelicans (large water birds belonging to family Pelecanidae, characterized by a long beak and a large throat pouch) eating anchovies (small, common saltwater forage fish of the family Engraulidae) off the California coast were dying from DA toxicity (Fig. 3.11). Other clinical symptoms are nausea, muscle weakness, disorientation and organ failure.

Formation of marine snow (aggregations of biological debris) in the surface waters by the toxic diatom *Pseudo-nitzschia australis* was experimentally shown by Schnetzer et al. (2017) when particulate and dissolved domoic acid (pDA and dDA) differed significantly among exponential phase. This was also occurred in deeper waters at 4 °C, and the authors ascertained that the diatom-derived marine snow potentially acts as a major vector for toxin flux to depth diatom.

Pseudo-nitzschia was the dominant diatom in Lim Bay, in the north-eastern Adriatic Sea, where a good assemblage of the species was recorded by Ljubešić et al. (2011) (Fig. 3.12). They have recorded *Pseudo-nitzschia manii* and potentially toxic *Pseudo-nitzschia pseudodelicatissima*, *Pseudo-nitzschia pungens*, *Pseudo-nitzschia fraudulenta* and *Pseudo-nitzschia calliantha* as the dominant species causing algal blooms. *Pseudo-nitzschia* abundance positively correlated to temperature,

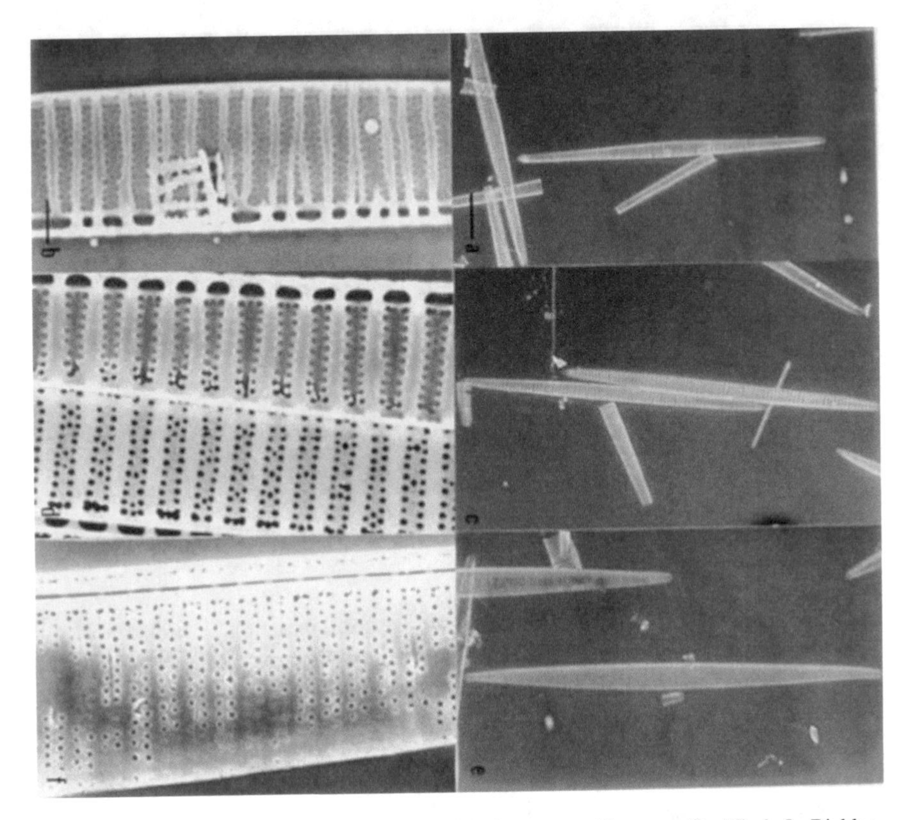

Fig. 3.12 *Pseudo-nitzschia* viewed through SEM microscope. (Courtesy: Dr. Mindy L. Richlen, National Office for HABs, WHOI, USA)

phosphate and ammonia in accordance with its maximal abundance in the summer/autumn period. Domoic acid was detected in the breeding population of the bivalve shellfish *M. galloprovincialis* but not certain which species is responsible. Based on 2-year intensive study, recently Bresnan et al. (2017) established the marked difference in domoic acid (DA) uptake and depuration in two shellfishes of commercial importance in Scotland, namely, blue mussels (*Mytilus edulis*) and king scallops (*Pecten maximus*), and these shellfishes might be used to act as a proxy for DA in the environment (Méndez et al. 2012). The Dirección Nacional de Recursos Acuáticos (DINARA) monitored a long-term programme along the Uruguayan coast and identified six species of *Pseudo-nitzschia*, namely, *P. australis*, *P. delicatissima*, *P. fraudulenta*, *P. multiseries*, *P. multistriata* and *P. pungens*. Méndez et al. (2012) recorded the most abundant *Pseudo-nitzschia* blooms (>20,000 cells l^{-1}) occur at salinity >30 in Uruguayan waters and confirmed the presence of potentially DA-producing Pseudo-*nitzschia* species in Uruguayan waters, namely, *P. fraudulenta*, *P. multiseries*, *P. multistriata* and *P. pungens*.

3.3.4 *Neurotoxic Shellfish Poisoning (NSP)*

Neurotoxic shellfish poisoning (NSP) is typically caused by ingesting bivalve shellfish (e.g. clams, oysters and mussels) that are contaminated with brevetoxins. Fish and bird mortality and contaminated shellfish (Deeds et al. 2010; van Deventer et al. 2012; Fauquier et al. 2013; Driggers et al. 2016), in addition to respiratory and gastrointestinal illnesses, were caused by brevetoxin exposures and neurotoxic shellfish poisoning (Fleming et al. 2011; Hoagland et al. 2014; Kirkpatrick et al. 2011; Pierce and Henry 2008; Reich et al. 2015; Steidinger 2009; Watkins et al. 2008; Ulloa et al. 2017). The risk for NSP toxins in shellfish is associated with HABs or 'red tides' along the Gulf of Mexico (heavy fish mortality). Harmful algal blooms and associated outbreaks of NSP have also been reported in New Zealand and Mexico (Ishida et al. 1996; Sim and Wilson 1997; Hernández-Becerril et al. 2007). The diagnosis of NSP is based upon clinical presentation and history of bivalve shellfish consumption from a risky area. Symptom onset may range from a few minutes to 18 h after consuming contaminated shellfish; however, in most cases, time to illness is about 3–4 h (Morris et al. 1991; Poli et al. 2000). The symptoms of NSP include both gastrointestinal and neurological problems. The most frequently reported symptoms are nausea, vomiting, abdominal pain and diarrhoea; however, these are not often the primary presenting complaint. Of greater concern to most individuals are the neurological symptoms which may include paraesthesia (abnormal tingling sensation) of the mouth, lips and tongue, peripheral tingling, partial limb paralysis, slurred speech, dizziness, ataxia (shaky movements resulting from brain failure to regulate), and a general loss of coordination. Reversal of hot/cold sensation, similar to ciguatera poisoning, has also been reported (Arnold 2011). Recent studies suggest that aerosolization of the toxin from seawater produces a transient, self-resolving, inhalational syndrome characterized by respiratory problems and eye irritation (Fleming et al. 2005, 2011). Exposure was associated with wave action and aerosolized sprays along the affected beaches during 'red tide' events. Adverse respiratory effects include upper airway irritation and discomfort, decreases in pulmonary function and exacerbation of symptoms in people with asthma.

3.3.5 *Ciguatera Fish Poisoning (CFP)*

CFP is a typical nonbacterial seafood-borne disease causing harmful effects on the people caused by the consumption of tropical coral reef fish contaminated with ciguatoxins (CTXs), polyether neurotoxins produced by the ciguatoxic dinoflagellates (*Gambierdiscus toxicus*). These are typically epiphytic to many types of seaweeds and appear to need nutrients exuded by these macroalgae. They flourish in areas of human or natural disturbance such as dredging, hurricanes and destruction of coral reefs. The 'new surface hypothesis' proposed by Randall in 1958 asserts that the multiple nature of human-induced and natural stresses (such as cyclones,

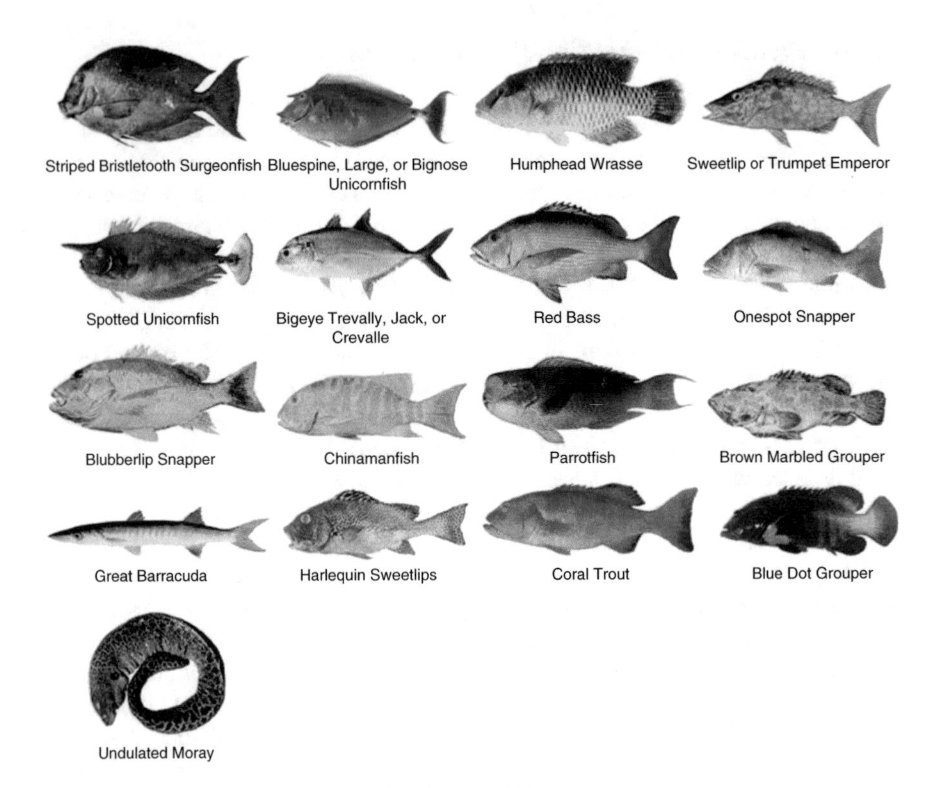

Fig. 3.13 Vector fish for Ciguatera Fish Poisoning (CFP)

tsunamis, coral bleaching, *Acanthaster planci* (the crown-of- thorns starfish) outbreaks, dredging, boat channel construction, boat anchorage and shipwrecks) provide freshly denuded surfaces for macroalgae to serve as substrate for toxic dinoflagellates (Cooper 1964; Banner 1976; Bagnis et al. 1988; Kohler and Kohler 1992; Bagnis 1994; Chinain et al. 1999).

Two groups of compounds implicated in the poisoning are ciguatoxin (CTX) and maitotoxin; both the toxin groups are produced by the *G. toxicus*. Ciguatoxins bioaccumulate in herbivorous fish, subsequently concentrate in large, carnivorous fishes through the marine food chain (Lewis 2006) (Fig. 3.13). However, other dinoflagellates have also been involved in CFP (Habekost et al. 1955; Dawson et al. 1955; Hashimoto et al. 1976; Laurent et al. 2008).

Reef fishes along with other invertebrates accumulate toxin produced by the ciguatoxic dinoflagellate and ciguatera poisoning affects at least 50,000–500,000 people per year worldwide (Fleming et al. 1998). CFP malady occurs in virtually all subtropical to tropical US waters (Hawaii, Florida, Guam, Puerto Rico, US Virgin Islands and other Pacific territories). The global economy has been highly affected hampering the fishery development around the world.

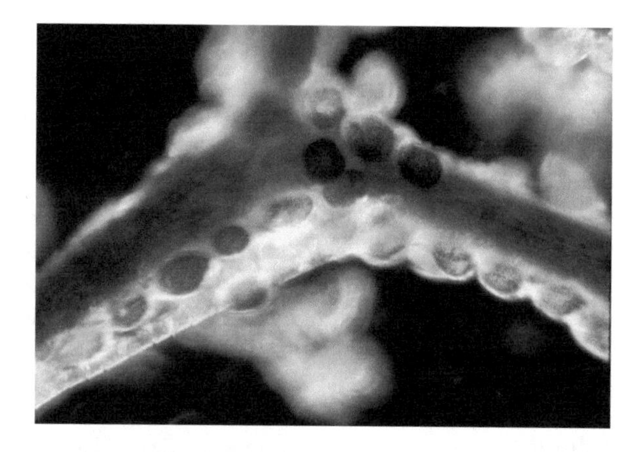

Fig. 3.14 *Gambierdiscus* cells on macroalgae. (Courtesy: Dr. Mindy L. Richlen, National Office for HABs, WHOI, USA)

Recently Rodríguez et al. (2017) have identified the Canary Islands (NE Atlantic) as a biodiversity 'hotspot' inhabited by five *Gambierdiscus* species (*G. australes*, *G. caribaeus*, *G. carolinianus*, *G. excentricus* and *G. silvae*), together with a new putative species (i.e. commonly accepted species) (*Gambierdiscus* ribotype 3) along 600 km longitudinal scale (Fig. 3.14). They have ascertained that some CFP cases in the region might be linked with the accumulation of ciguatoxins (polyether marine toxins) in the marine food web acquired from local populations of *Gambierdiscus*. Warmer climate conditions might be conducive for such unique congeneric diversity existing in the Miocene Epoch (when oldest current Canary Islands were created), in contrast with cooler present ones. Currently, warming trends associated with climate change could also act to extend the favourable environmental set-up in the area for growth of the species especially during winter months.

Rongo and Woesik (2012) examined the socioeconomic consequences of CFP on small island communities in Rarotonga, Southern Cook Islands. This is revealed that CFP halved the per capita fresh fish consumption during 1989–2006 and the economic consequences of ciguatera amounted to around NZD $750,000 per year and may have changed the social, cultural and traditional characteristics of the prior subsistence fishing lifestyle.

Clinical symptoms: In humans, ciguatera poisoning occurs after ingesting the toxin. The symptoms include acute gastrointestinal (nausea, diarrhoea, vomiting), neurological (reversal of thermal sensation) and cardiovascular disorders (low blood pressure) and bradycardia (slowing of heart rate) and in extreme cases death. Other symptoms include joint pain, miosis (constriction of pupil) and cyanosis bluish discolouration of skin and mucous membrane). In addition, some studies have linked outbreaks of cyanobacteria to ciguatera poisoning-like symptoms.

3.3.6 Azaspiracid Shellfish Poisoning (AZA)

Azaspiracid (AZA) poisoning, a human toxic syndrome caused by a new set of toxins associated with shellfish consumption via food web, is etiologically similar to diarrheic shellfish poisoning (DSP). Similar to DSP toxins, human consumption of AZA-contaminated shellfish caused illness manifesting by chills, headaches, nausea, vomiting, diarrhoea, abdominal heaviness and stomach cramps.

The severe gastrointestinal illness in humans is caused due to consumption of bivalve shellfish species (*Mytilus edulis*), which has nutritive value including high levels of polyunsaturated fatty acids (PUFA). However, azaspiracid is believed to be far more toxic than okadaic acid, the primary cause of DSP. In 1995 the first report of AZP came from the Netherlands where blue mussels (*Mytilus edulis*) originating from Killary Harbour, Ireland, induced DSP-like symptoms in humans. Similar cases of AZP were also reported in Ireland (1997), France (1998) and Italy (1998). The contaminated shellfish causing these illnesses were cultivated in four different regions encompassing the entire west coast of Ireland. Toxins have also been identified in mussels throughout most of Western Europe as well as north-western Africa and Canada which implies a more widespread intoxication of shellfish than was previously thought.

AZAs are a group of lipophilic phycotoxins, synthesized by some species of the marine planktonic Dinophyceae genus *Azadinium*, and accumulate in a number of shellfish species such as oysters, clams and scallops (Furey et al. 2002, 2003; Magdalena et al., 2003b) along with a decapods crustacean (crab) from Norway (Torgersen et al. 2008). These are comprised of novel structural features, where the prevalent AZAs are AZA1, AZA2 and AZA3 which differ from each other in their degree of methylation. Later on, two more analogous of AZA were identified in mussels, namely, AZA 4 and AZA 5.

AZAs have been first discovered in Ireland but are now reported in shellfish from numerous global sites thus showing a wide distribution. *Azadinium poporum* Tillmann & Elbrächter is a small dinoflagellate from the family Amphidomataceae which is known for the potential production of azaspiracids (AZAs) causative of azaspiracid shellfish poisoning (AZP). Recently, Tillmann et al. (2017) first recorded three strains of *Azadinium poporum* from the Chilean coastal regions, Pacific side of South America, endorsing the risk of AZA shellfish and concomitant human contamination in these regions. Kim et al. (2017) also identified the *A. poporum* and a new azaspiracid (AZA) from this species in Puget Sound, Washington State, USA. Strikingly, AZA could not be detected in any strains of *Azadinium obesum*, *A. cuneatum* and *A. dalianense*, but all four strains of *A. poporum* produced a new azaspiracid toxin (AZA-59), based on LC-MS analysis. Akselman and Negri (2012) recorded two successive blooms *A.* cf. *spinosum* in northern shelf waters of Argentina, Southwestern Atlantic, occurred extensive and highly productive areas during spring in successive years. They have asserted that the dinoflagellate was the most important species in middle shelf waters, with high-biomass and abundance values. The presence of these AZA-producing species deserves special attention in

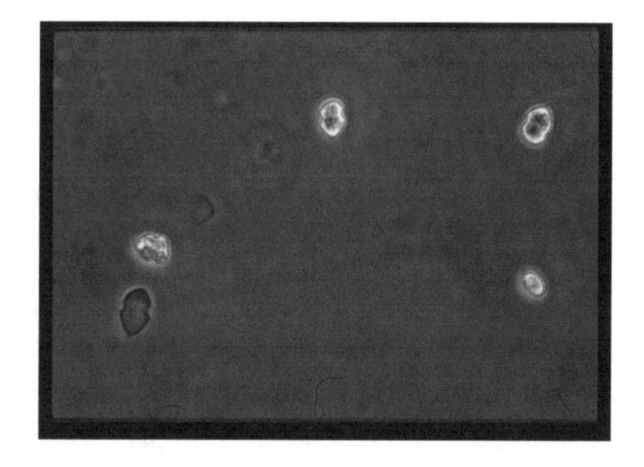

Fig. 3.15 *Azadinium* cells (in light microscope) from the 1991 bloom in the Argentine Sea, Southwestern Atlantic. (Courtesy: Dr. RutAkselman, InstitutoNacional de Investigación y DesarrolloPesquero, Argentina)

the context of bivalve commercial fisheries. Figure 3.15 shows *Azadinium* cells (in light microscope) from the 1991 bloom in the Argentine Sea, Southwestern Atlantic.

AZAs have been recorded in shellfish harvested throughout Europe (James et al. 2000, 2002, 2004, 2008; Magdalena et al. 2003a; Vale 2004; Twiner et al. 2008; Vale et al. 2008), Northeast Africa (Taleb et al. 2006); Southwestern Atlantic (Akselman and Negri 2012); Puget Sound, USA (Kim et al. 2017); southwest Pacific Region (Tillmann et al. 2017) and also Norway (TorgerSen et al. 2008). Aerial photo of the spring bloom of 1990 38° 50′S -56° 10′W (Fig. 3.16) shows the stele of fishing vessels when water was stirred from subsurface levels in which there was higher cell densities; this was detected in the 1990s decade (1990, 1991, 1998) which occurred in the Argentine Sea, Southwestern Atlantic.

Based on analysis using LC-MS/MS, Tillmann et al. (2017) established the presence of two different AZA analogues in several bivalve shellfish species (*Chamelea gallina*, *Cerastoderma edule*, *Donax trunculus*, and *Solen vagina*) at the Atlantic coast of southern Spain. AZA levels exceeded the EU regulatory level of 160 μg AZA-1 eq. kg^{-1} in majority of the samples studied (reaching maximum levels of >500 μg AZA-1 eq. kg^{-1} in *Chamelea gallina* and > 250 μg AZA-1 eq. kg^{-1} in *Donax trunculus*). The contamination status highlighted the risk of azaspiracid poisoning (AZP) for this area as well as also for the Atlantic coast of Iberia and North Africa. Azaspiracids are a suite of toxins which are more dangerous and significantly different from other classes of lipophilic shellfish toxins, and these accumulate in bivalve molluscs through feeding on toxic microalgae. Widespread occurrence of AZP is very much evident, and this can be minimized by adopting the following measures: (1) a combination of molecular methods for species detection and solid phase resin deployment to target shellfish monitoring of toxin (Kim et al. 2017), (2) proper risk assessment and regulatory control of shellfish and (3) high level of vigilance to limit human exposure to these seafood toxins.

The above discussion strongly suggests the need for continuous monitoring of AZA and the organisms producing such toxins. Liquid chromatography hyphenated with tandem mass spectrometry (LC-MS/MS) is perhaps the most effective means

Fig. 3.16 Aerial photo of the spring bloom of 38° 50′S -56° 10′W which shows the stele of fishing vessels when water was stirred from subsurface levels in which there was higher cell densities; this was detected in the 1990 decade (1990, 1991, 1998) occurred in the Argentine Sea, Southwestern Atlantic. (Courtesy: Dr. Hugo Benavides, InstitutoNacional de Investigación y DesarrolloPesquero, Argentina)

of AZA determination in shellfish. This is based on the assumption that 'AZA' represents a group of structurally similar polyethers, with different toxicologies and without an analytically discernible chromophore, and is present at trace amounts amidst the very complicated matrix of shellfish, often in conjunction with other shellfish toxins, including DSP and spirolides (marine phycotoxins produced by the dinoflagellates) (James et al. 2004; Álvarez et al. 2010).

3.3.7 Cyanobacterial Harmful Algal Bloom (cyanoHAB)

Blooms associated with photosynthetic prokaryotes (cyanobacterial harmful algal blooms [cyanoHAB]) seemed to be increasing globally over the past few decades, but relatively little quantitative information is available about the spatial extent of blooms. To assess cyanoHAB risk, the World Health Organization's (WHO) recreational guidance level thresholds have categorized the surface area of cyanoHABs into three risk categories: low-, moderate- and high-risk bloom area. Recently, Urquhart et al. (2017) developed an assessment method using Medium Resolution

Imaging Spectrometer (MERIS) imagery to quantify cyanoHAB surface area extent, transferable to different spatial areas.

Cyanobacterial blooms have been increasing worldwide due to increased nutrients associated with urban, industrial and agricultural development. Blooms that occur in the Indian River Lagoon (IRL), Florida, may be increased by nutrient-laden runoff from storm water and non-point sewage pollution due to alterations to the watershed. In the IRL, during the summer of 2006, extensive blooms of the marine cyanobacterium, *Lyngbya majuscula*, were observed forming mats throughout beds of the seagrass Halodule *wrightii* in Fort Pierce, Florida (Tiling and Proffitt 2017). The effects of cyanobacterial blooms were compared to artificial shading of *H. wrightii* to assess the shading potential of *L. majuscula*. The combined effects of *L. majuscula* removal and artificial shading showed increases in the below ground biomass of *H. wrightii*. However, leaf length increased in the presence of *L. majuscula*. In response to artificial shading, *H. wrightii* decreased in density but showed similar leaf elongation. A common bivalve mollusc, *Macoma constricta,* increased in density when *L. majuscula* was removed. Therefore, when *L. majuscula* blooms occur, light limitation is one of the mechanisms altering *H. wrightii* density and leaf lengths in the IRL. Loss of *H. wrightii* biomass due to shading from cyanobacterial mats may further damage the diversity and habitat value of the IRL.

Cyanobacteria exhibit rapid growth under high-nutrient conditions and as such have been increasingly referred to as harmful algal blooms (HABs) when they become excessive in aquatic ecosystems. While nutrient runoff is thought to be the primary cause of HABs, it has been found that it is not necessarily the independent quantities of nitrogen and phosphorus that favour cyanobacteria but rather the N:P ratio. However, in contrast, cell growth and paralytic shellfish poison (PSP) content were related on N:P ratio in the dinoflagellate *Alexandrium tamarense* (Murata et al. 2012), which has been discussed in Chap. 3.

Cyanobacteria are aquatic and photosynthetic, that is, they live in the water, and can manufacture their own food. Because they are bacteria, they are quite small and usually unicellular, though they often grow in colonies or filaments, often surrounded by a gelatinous or mucilaginous sheath. They have the distinction of being the oldest known fossils, more than 3.5 billion years old, in fact. It may surprise you then to know that the cyanobacteria are still around; they are one of the largest and most important groups of bacteria on earth. Many Proterozoic oil deposits are attributed to the activity of cyanobacteria. They are also important providers of nitrogen fertilizer in the cultivation of rice and beans. The cyanobacteria have also been tremendously important in shaping the course of evolution and ecological change throughout earth's history.

Cyanobacteria are very important organisms for the health and growth of many plants. They are one of very few groups of organisms that can convert inert atmospheric nitrogen into an organic form, such as nitrate or ammonia. It is these 'fixed' forms of nitrogen which plants need for their growth and must obtain from the soil. Fertilizers work the way they do in part because they contain additional fixed nitrogen which plants can then absorb through their roots. Nitrification cannot occur in the presence of oxygen, so nitrogen is fixed in specialized cells called heterocysts.

These cells have an especially thickened wall that contains an anaerobic environment. Many plants, especially legumes, have formed symbiotic relations with nitrifying bacteria, providing specialized tissues in their roots or stems to house the bacteria, in return for organic nitrogen. This has been used to great advantage in the cultivation of rice, where the floating fern `Azolla` is actively distributed among the rice paddies. The fern houses colonies of the cyanobacterium `Anabaena` in its leaves, where it fixes nitrogen. The ferns then provide an inexpensive natural fertilizer and nitrogen source for the rice plants when they die at the end of the season. Cyanobacteria also form symbiotic relationships with many fungi, forming complex symbiotic 'organisms' known as lichens.

3.4 Possible Detection Measures for HABs

Selective marine pennate diatoms of the genus *Pseudo-nitzschia* are concerned with the production of domoic acid, a naturally occurring heterocyclic amino acid responsible for amnesic shellfish poisoning. This is somehow problematic and time-consuming to differentiate between the potentially toxic and non-toxic species belonging to this genus. This requires a detailed study with the help of scanning electron microscopy (SEM) of cleaned frustules. The whole-cell (in situ) hybridization and species-specific large subunit ribosomal RNA (LSU rRNA)-targeted oligonucleotide probes are the most convenient option for identification of these species, as demonstrated by Miller and Scholin (1996). The technique is a rapid, simple and cost-effective for discriminating among *Pseudo-nitzschia* species, hence helpful for adopting optimal management strategy as it affords early warning of harmful algal blooms.

The same technique was also applied by Touzet and Raine (2007) to accurately discriminate between the two species of *Alexandrium* species, namely, *A. andersoni* and *A. minutum*, which is time and cost-effective and definitely requires the help of taxonomic expert as the genus contains morphologically similar toxic and non-toxic species. As previously discussed, the species are responsible for paralytic shellfish poisoning (PSP) and hence require the reliable identification and enumeration of vegetative stages in order to enable the development of early warning policies. In general, *Alexandrium* spp. are identified based on the characteristics of their morphological features with the help of conventional light microscopy. However, the drawback is that one cannot identify at the species level and does not allow their characterization at the species level. The development and application of large subunit (LSU) rRNA-targeted oligonucleotide probes for the detection and quantification of *A. minutum* (Global clade) and *A. andersoni* by whole-cell fluorescent in situ hybridization (FISH) have been considered the best option for correct, globally consistent and rapid identification of target species.

A considerable effort has been made over the past decade to develop species-specific oligonucleotide probes for correct, globally consistent and rapid identification of target species (e.g. Miller and Scholin 1998). This is an optimal management strategy as it affords early warning of potentially toxic blooms. However, unanticipated interspecific diversity in sequences among toxicogenic diatoms precluded the use of certain probes in regions geographically distinct from the source of the original target strains (Parsons et al. 1999; Turrell et al. 2008). This underscores the need for in-depth investigations of both local and global diversity of the genus *Pseudo-nitzschia*. A powerful approach to addressing the challenges posed by interspecific diversity is to construct clone libraries for a given area using genus- specific large subunit (LSU) rDNA primers for polymerase chain reaction-based amplification of environmental samples (e.g. McDonald et al. 2007), which can provide a qualitative assessment of genotypic diversity within a sample and also guide development of probes for detection of semi-cryptic species. Rapid advances are also being reported in the application of DNA probe and toxin detection technologies by remote, subsurface platforms capable of autonomously providing near real-time data on *Pseudo-nitzschia* spp. and DA concentrations (Scholin et al. 2009).

For a reliable assessment and management of the risks linked to CFP causing dinoflagellate *Gambierdiscus* sp., proliferation, Roué et al. (2018) first demonstrated the suitability of adsorption toxin tracking solid phase adsorption toxin tracking (SPATT) technology for in situ monitoring of CTXs. For such passive monitoring devices, they had used porous synthetic resins capable of adsorbing toxins directly from the water column to assess ciguatera risk assessment.

Again, Dabrowski et al. (2016)developed a 3D primitive equation coastal ocean model which is deemed suitable for incorporation into the HAB warning system. This forecasting model was implemented for Southwest Ireland (Bantry Bay). Again, the 'Applied Simulations and Integrated Modelling for the Understanding of Harmful Algal Blooms' (Asimuth) project sought to develop a HAB alert system for Atlantic Europe (Davidson et al. 2016).

Algal bloom is the potential medium in the carbon sequestration process, but access to collect consistent data on them is not always plausible. Considering the present restrictions, the remOcean 3 project (remotely sensed biogeochemical cycles in the ocean), led by LOV researchers, developed the biogeochemical profiling float as a supplementary observation technique to measure temperature and salinity levels up to 2 km depth. The vertical movement of these battery-run robotic devices is controlled by a hydraulic system that inflates or deflates a buoy-like bladder by pumping oil into or out of it. Data collected by the float's sensors is transferred to scientists by a satellite antenna, offering 'a two-way communication system that also allows researchers to send instructions to the robot'.

3.5 Modified Clay for Controlling of HABs

Application of natural clays (the fine-grained particles; <0.0002 mm), the basic component of soil, is considered the most ecofriendly approach for controlling the HABs in a large scale. So far, many methods have been adopted, either chemical or biological, for killing the HAB organisms as well as inhibiting bloom formation. However, clay application seems to be superior than the other due to less pollution and ecological impact, low cost and most conventional in using in the field. Hence the clay disposal method gathered increasing interest for the last several years (Anderson 1997; Kim 2006; Park et al. 2013; Sengco and Anderson 2004; Sengco et al. 2001; Yu et al. 1994a). The mechanism related to this mitigation processes related to flocculation and/or entrainment and accelerated sedimentation of algal cells through ecologically inert clay application and should be better recognized as clay-algal coupling. However, these natural clays have low flocculation rate, and thus a huge clay amount is required to get an effective efficiency in the field which is cost-effective, and ecological impact is also high. Again, organic enrichments of sediments might also result due to repeated application of clay to control the bloom activities. Hence clays have limited application due to these negative ecological impacts, and further research was carried out by many scientists (Lee et al. 2008; Liu et al. 2010; Maruyama et al. 1987; Miao et al. 2014; Sengco et al. 2001; Sun et al. 2004a, b; Yu et al. 1999) to improve and upgrade the removal efficiency of clay. In the 1990s, the interactions between the clay particles and HAB organisms were intensely studied by Yu and co-workers (see Yu and Zou 1994; Yu et al. 1994a, b, c, 1995) who determined the key factors controlling the flocculation efficiency of clays.

Various modified clays composed of different organic compounds have been prepared, the removal efficiencies of which are multiple times greater than unmodified natural clays. MCs are promising effective controlling option through rapid sedimentation of HABS through flocculation, hence widely used for the mitigation of HABs caused by diverse species, as described below:

(a) Zhang et al. (2018) demonstrated the useful application of modified clay (MC) in appropriate concentrations for the control of HABs on cyst formation and germination of *Alexandrium pacificum*. MC might be considered as a promising ecofriendly approach for effective management of *A. pacificum* blooms because it does not leave more residual cysts and most importantly MC has high efficiency and low environmental impacts.
(b) Liu et al. (2010) modified the clays collected from Lake Taihu, China, with hexadecyl trimethyl ammonium bromide (CTAB) and applied to clean and control the *Microcystis aeruginosa* blooms in lake water. The CTAB-modified clays showed unique potential in inhibiting >90% mobility of algal cells and addition of CTAB largely enhanced the clay effects.
(c) Lu et al. (2015) highlighted the environmental effects of modified clay (MC) flocculation on toxic dinoflagellate *Alexandrium tamarense* releasing paralytic shellfish poisoning toxins and observed that MC quickly eliminated inorganic and organic macronutrients from seawater. In addition, detoxification of PSTS by using MC treatment was also confirmed the transformation of high toxicity and gonyautoxins 1 and 4 (GTX1 and GTX4) to the lower toxicity decarbamoyl gonyautoxins (dcGTX3) and gonyautoxin 2 (GTX2).
(d) Liu et al. (2016a, b) compared the efficiencies of aluminium chloride-modified clay (AC-MC), aluminium sulphate-modified clay (AS-MC) and polyaluminium chloride-modified clay (PAC-MC) in removing *Aureococcus anophageffe-*

rens. The results revealed that AC-MC and AS-MC had better efficiencies in removing the algal species than PAC-MC and the process was chiefly governed by surface charge at the pH of the algal culture after addition of clay.

(e) Ma and Liu (2002) performed laboratory studies to assess the effectiveness of potassium ferrate (VI) preoxidation on removal of algae by coagulation with alum ($Al_2(SO_4)_3 \cdot 18H_2O$), and it was revealed that a very short time was good enough to achieve substantial algal removal efficiency. It was also observed that this kind of pretreatment enhances the reduction of alum dosage required to cause an efficient coagulation for algae removal.

3.6 Impact of Climate Change on the Occurrence of HABs

The occurrence of red tides and HABs, with increasing frequency, intensity and impacts worldwide as a result of steady changes in ocean climate, increased enrichment of anthropogenic nutrients (eutrophication) and enhanced long-distance dispersal in ballast water (Landsberg 2002; Heisler et al. 2008; Rabalais et al. 2009; Griffith and Gobler 2016). The increased incidence of harmful algal blooms can adversely affect organisms from micro-scale to ecosystem scale (Landsberg 2002; Graneli and Turner 2008).

As a consequence of climate change, the sea surface temperature (SST) has been on the rise, and coastal upwelling events have been more and more frequent (as discussed in Chap. 2). The potential impacts of climate variability on interannual fluctuations on the community structure and numerical abundance of diverse functional groups and trophic levels of plankton have received great attention in recent decades (Fromentin and Planque 1996; Planque and Taylor 1998; Edwards and Richardson 2004).

The incidence of HABs caused by dinoflagellates may be increasing as a consequence of climate change. The southern Benguela Current Large Marine Ecosystem (LME) has experienced extensive dinoflagellate blooms in recent years. The Benguela sardine or South American pilchard (*Sardinops sagax*) (Clupeidae) is an economically and ecologically important small pelagic oceanodromous fish which suffered concomitant side effects (e.g. feeding) due to occurrence of dinoflagellate bloom *Gonyaulax polygramma*, the non-toxic red tide bloom species. Van der Lingen et al. (2016) hypothesized that the species had negative impact on the fish as it possesses a fine-meshed branchial basket that can retain the dinoflagellates.

A high abundance of resting cysts of *Alexandrium tamarense* was observed recently by Natsuike et al. (2017) in the vast continental shelf of the Chukchi Sea, marginal sea in the Arctic Ocean, endorsing its occurrence in the shelf. Sea ice reduction and the inflow of Pacific summer water (PSW) through the Bering Strait have recently increased due to warming in the shelf region. The species occurrences were mainly recorded specifically PSW-affected sites, especially Bering shelf water. Water structure of PSW was characterized by warmer surface and bottom water temperatures, and increased temperatures may have promoted the cell growth and

Fig. 3.17 Mass mortality of *Alexandrium* clams in Africa due to occurrence of HABs. (Courtesy: Dr. Mindy L. Richlen, National Office for HABs, WHOI, USA)

Fig. 3.18 Massive fish kill resulting from anoxia caused by *Ceratium* sp. in South Africa. (Courtesy: Dr. Mindy L. Richlen, National Office for HABs, WHOI, USA)

cyst germination of *A. tamarense*. An increase in the PSW inflow due to warming favours occurrences of this toxic species on the Chukchi Sea shelf. Mortality of *Alexandrium* clams in Africa due to occurrence of HABs (Fig. 3.17) and mass fish kill resulting from anoxia caused by *Ceratium* sp. in South Africa (Fig. 3.18) were the two major impacts on World Ocean.

Fig. 3.19 Mortality of whale due to negative impact of HABs. (Courtesy: Dr. Mindy L. Richlen, National Office for HABs, WHOI, USA)

Unusual mortality in bottlenose dolphin (*Tursiops truncatus,* Montagu, 1821; family Delphinidae) due to co-occurrence of multiple classes of HABs caused by toxin producing genera (*Dinophysis* sp., *Prorocentrum* sp. and *Pseudo-nitzschia* sp.) was recorded by Fire et al. (2011) in Texas coastal waters. *T. truncatus* is the most well-known aquatic mammals inhabiting in warm and temperate seas worldwide. Nutrient pollution or enrichment is now a major agent of global change in coastal water linking an array of coastal problems including eutrophication, biodiversity loss of seagrass and coral reef ecosystem (NRC 2006; Rockström et al. 2009). With the apparent increase of frequency and intensity of HABs around the globe, more attention should be paid to the risks of seafood contamination resulting human intoxication and socioeconomic impacts. The mortality of two vulnerable marine mammals, such as whale (Class Cetacea) (Fig. 3.19) as well as Dugongs ('sea cow' order Sirenia) (Fig. 3.20), is also associated with the negative impact of HABs.

Dinophysis acuta (Ehrenberg), one of the main agents of diarrhetic shellfish poisoning (DSP), is a frequent seasonal lipophilic toxin producer in European Atlantic coastal waters associated with thermal stratification (Díaz et al. 2016). In the Galician Rı'as, populations of *D. acuta* with their epicentre located off Aveiro (Northern Portugal) typically co-occur with *Dinophysis acuminata* (Clapere'de & Lachmann) during the upwelling transition (early autumn) as a result of longshore transport. During hotter than average summers, *D. acuta* blooms also occur in August in the Rı'as, when they replace *D. acuminata*. A prolonged time series during 1985–2014 of *D. acuta* in the Galician Rı' was to identify the distribution patterns and their rela-

Fig. 3.20 Dugongs (manatee): the vulnerable marine mammal in which mortality was recorded due to impact of HABs. (Courtesy: Dr. Mindy L. Richlen, National Office for HABs, WHOI, USA)

tion with climate variability, and also to explain the exceptional summer blooms of the species in 1989–1990. A dome-shaped relationship was found between summer upwelling intensity and *D. acuta* blooms; cell maxima were associated with conditions where the balance between upwelling intensity and heating, leading to deepened thermoclines, combined with tidal phase (3 days after neap tides) created windows of opportunity for this species. The application of a generalized additive model based on biological (*D. acuta* inoculum) and environmental predictors (cumulative June–August upwelling CUIJJA, average June–August SSTJJA and tidal range) explained more than 70% of the deviance for the exceptional summer blooms of *D. acuta*, through a combination of moderate (35,000–50,000 m^3 s^{-1} km $^{-1}$) summer upwelling (CUIJJA), thermal stratification (SSTJJA >17 8C) and moderate tidal range (2.5 m), provided *D. acuta* cells (inoculum) were present in July. There was no evidence of increasing trends in *D. acuta* bloom frequency/intensity nor a clear relationship with NAO or other long-term climatic cycles. Instead, the exceptional summer blooms of 1989–1990 appeared linked to extreme hydroclimatic anomalies (high-positive anomalies in SST and NAO index), which affected most of the European Atlantic coast.

Studies of cyclone-induced blooms have focused on changes in the universal algal photopigment chlorophyll a (Chl a) to indicate algal production (Walker et al. 2005; Zheng and Tang 2007; Reddy et al. 2008; Paerl et al. 2001a, b; Miller et al. 2006). Weather-related disturbances, such as wind-generated waves, major rainfall events and large temperature shifts associated with frontal passages, are the key drivers of ecological processes in shallow lakes (Zhu et al. 2014). The influence of Pacific tropical cyclones on cyanobacterial blooms in China's third largest lake, Taihu, was studied during the passage of two typhoons using a continuous monitoring as part of an on-lake high-frequency recording platform, coupled to satellite-based remote sensing data. Short-term (on the order of hours) nutrient pulsing resulting from the passage of typhoons played a key role in bloom initiation and maintenance. Decreasing wind speeds and increasing air and water temperatures in the aftermath of cyclones were accompanied by enriched algal biomass concentrations. The synergistic effects of nutrient pulsing, elevated water temperatures and increased water column stratification after the passage of the cyclones stimulated blooms of the toxic cyanobacteria *Microcystis* spp. There were short-term successions of blooms following typhoons, and as blooms 'crashed' they provided nutrient inocula for future blooms. Trends determined from historical in situ data indicated higher frequencies and intensities of blooms in 'cyclone years'. Typhoons are an important driver of biogeochemical and water quality perturbations at the ecosystem level in this hypertrophic lake.

During the summer of 2013, about 70 people received diarrhetic shellfish poisoning following consumption of mussels harvested in the Shetland Islands, Scotland (Whyte et al. 2014). At this time, large numbers of the biotoxin-producing phytoplankton genus *Dinophysis* were observed around the Shetland Islands. Analysis indicated this increase was not due to in situ growth but coincided with a change in the prevalent wind direction. A previous large bloom of *Dinophysis* during 2006 also coincided with a similar change in the prevalent wind patterns. Wind direction and speed in the North East Atlantic and the North Sea is strongly influenced by the North Atlantic oscillation (NAO) with a positive relationship between it and wind direction. A positive trend in the NAO was linked to climate change and predictions suggested that there would be an increasingly westward component to prevalent wind directions in the North Sea which could lead to an increase in the occurrence of these harmful algal blooms. Wind patterns analyses, therefore, seem to be one of the potential methods of early warning of future bio-toxicity events.

Parasites of phytoplankton influence phytoplankton bloom dynamics and may severely affect the type of food available for higher trophic levels. The incidence of parasitic infections generally seemed to increase across ecosystems worldwide under global change impact. Recently, Guinder et al. (2017) reported a massive parasite infection on two dominant diatoms of the austral winter bloom, namely, *Thalassiosira pacifica* Gran and Angst, 1931, and *Chaetoceros diadema* (Ehrenberg) Gran, 1897. These diatoms were recorded during an extreme precipitation period in the Bahía Blanca Estuary, Argentina. The parasite infection was coincided with a drastic fall in water salinity and affected >40% of host cells. The possible parasite on *T. pacifica* was mostly similar to *Pirsonia* sp., a nanoflagellate with high host specificity. After the intense rainy period and the parasitic infection, the phytoplankton biomass drastically dropped (~ 80%), and the community structure shifted to one with smaller species (i.e. *Thalassiosira curviseriata*, *T. hibernalis* and *T. minima*). The authors demonstrated that these modifications may have negative impact on the food web dynamics and the potential relationship between precipitation-driven modifications in water properties and the emergence of ephemeral parasitism in coastal eutrophic environments.

The effect of eutrophication and ocean acidification on dimethyl sulphide (DMS) emissions in eutrophied coastal environments is an important aspect in the context of climate change. As previously discussed in Chap. 2, coastal environments are 'hotspots' of DMS emission and are subject to eutrophication. It is to be noted that dimethylsulphoniopropionate (DMSP) synthesized by phytoplankton is the principal precursor of the climatically active gas DMS and the production of DMSP by phytoplankton ca be modified by nutrient availability.

3.7 Conclusion

Harmful algal blooms (HABs) have become a global marine disaster that can cause massive fish kills, contamination of sea food with toxins and ecological damage through the development of anoxia. The HABs prevalence is significantly growing worldwide with the intensification of anthropogenic activities, posing a major threat to human and ecosystem health, fisheries, coastal economics as well as safety of many industries using ambient seawater, such as nuclear power plants. The ecologically and economically disruptive HABs have expanded remarkably worldwide, leading a major threat to public as well as ecosystem health and to fisheries and coastal economics. It is evident from the above discussion that HAB event has massive negative impact by altering ecosystems through deoxygen (hypoxia/anoxia), stimulation of pathogenic bacteria, disrupt productivity and massive localized mortality (largely fish, marine mammals, including the endangered species (e.g., California sea otter), and humans). To explore the key factors that trigger and develop toxic/non-toxic algal blooms in dynamic coastal regions is critically complex. There is an urgent need to improve our capacity to predict future bloom events. A collaborative and comprehensive long-term monitoring programme is absolutely important to improve our understanding of nutrient and algal dynamics as a whole. To minimize human and health risks, extensive long-term collaborative monitoring and transdisciplinary research over the years are absolutely important involving the HAB researchers, local health departments at academy community outreach and policy levels.

HABs are complex natural phenomena that require multidisciplinary study ranging from molecular and cell biology to large-scale field surveys, numerical modelling and remote sensing from space. We should identify the major knowledge gaps regarding the understanding of bloom formation, detection and mitigation along with current bloom remediation efforts around the world. Many affluent countries have extensive seafood monitoring programmes to ensure toxin-free shellfish.

The diversity in HAB species and their impacts presents a significant challenge to those responsible for the management of coastal resources. Management and control strategies should include both precautionary and emergency aspects. Precautionary management includes establishing an intricate network for observation and prediction system, an early warning system and reducing damage to mariculture farms. Besides regular HAB monitoring activities, automatic HAB alarm system provided with turbidity and Chl a sensors should also be implemented simultaneously to predict HAB occurrence. Emergency management is required after a HAB outbreak to prevent damage to fisheries which includes transferring fish to a safe region, oxygen supply to fish, stopping feeding and clay dispersal. Clay is a natural component and non-toxic, inexpensive and easy to use in field operation. Hence proportionate use of clay might be the most ecofriendly option to control HABs.

HABs provide stakeholders and managers quantifiable cyanoHAB rates of change and spatial extent. An interagency HABHRCA (Hypoxia Research Control Act report in 2016) identified monitoring challenges, which included sustaining monitoring programmes and maintaining consistency of methods across monitoring programmes, a consistent approach for determining the spatial extent and rate of change, year-to-year, with long-term operational satellites.

References

Adam, A., Mohammad-Noor, N., Anton, A., Saleh, E., Saad, S., & Shaleh, S. R. M. (2011). Temporal and spatial distribution of harmful algal bloom (HAB) species in coastal waters of Kota Kinabalu, Sabah, Malaysia. *Harmful Algae, 10*(5), 495–502.

Ahn, Y. H., & Shanmugam, P. (2006). Detecting the red tide algal blooms from satellite ocean color observations in optically-complex Northeast-Asia coastal waters. *Remote Sensing of Environment, 103*, 419–437.

Aiyar, R. R. (1936). Mortality of fish of the Madras coast in June 1935. *Current Science, 4*(7), 488–489.

Akselman, R., & Negri, A. (2012). Blooms of *Azadinium* cf. *spinosum* Elbrächter et Tillmann (Dinophyceae) in northern shelf waters of Argentina, Southwestern Atlantic. *Harmful Algae, 19*, 30–38.

Al-Azri, A. R., Piontkovski, S. A., Al-Hashmi, K. A., Goes, J. I., Gomes, H. D. R., & Glibert, P. M. (2014). Mesoscale and nutrient conditions associated with the massive 2008 Cochlodinium polykrikoides bloom in the sea of Oman/Arabian Gulf. *Estuaries and Coasts, 37*(2), 325–338.

Álvarez, G., Uribe, E., Ávalos, P., Mariño, C., & Blanco, J. (2010). First identification of azaspiracid and spirolides in Mesodesma donacium and Mulinia edulis from Northern Chile. *Toxicon, 55*(2–3), 638–641.

Anderson, D. M. (1984). Shellfish toxicity and dormant cysts in toxic dinoflagellates blooms. In E. P. Ragelis (Ed.), *Seafood toxins*. American Chemical Society. Symposium 505 series N? 262, Washington, DC, pp. 125–138.

Anderson, D. M. (1997). Turning back the harmful red tide. *Nature, 388*(6642), 513–514.

Anderson, D. M., & Keafer, B. A. (1987). An endogenous annual clock in the toxic marine dinoflagellate *Gonyaulax tamarensis*. *Nature, 325*, 616–617.

Anderson, D. M., Aubrey, G., Tyler, M. A., & Coats, D. W. (1982). Vertical and horizontal distributions of dinoflagellate cysts in sediments. *Limnology and Oceanography, 27*, 757–765.

Anderson, D. M., Glibert, P. M., & Burkholder, J. M. (2002). Harmful algal blooms and eutrophication: Nutrient sources, composition, and consequences. *Estuaries, 25*, 704–726.

Anderson, D. M., Stock, C., Keafer, B., Nelson, A., McGillicuddy, D., Keller, M., Thompson, B., Matrai, P., & Martin, J. (2005). *Alexandrium fundyense* Cyst dynamics in the Gulf of Maine. *Deep Sea Research Part II, 52*, 2522–2542.

Anderson, D. M., Alpermann, T. J., Cembella, A., Collos, Y., Masseret, E., Montresor, M., & Truby, E. (2012). The globally distributed genus *Alexandrium*: Multifaceted roles in marine ecosystems and impacts on human health. *Harmful Algae, 14*, 10–35.

Arnold, T. C. (2011). *Shellfish toxicity. Medscape reference: Drugs, diseases & procedures*. http://www.emedicine.com/EMERG/topic528.htm

Backer, L. C., & McGillicuddy, D. J., Jr. (2006). Harmful algal blooms at the interface between coastal oceanography and human health. *Oceanography, 19*(2), 94–106.

Bagnis, R. (1994). Natural versus anthropogenic disturbances to coral reefs: Comparison in epidemiological patterns of ciguatera. *Memoirs of the Queensland Museum, 34*, 455–460.

Bagnis, R., Bennett, J., Barsinas, M., Drollet, J. H., Jacquet, G., Lecrand, A. M., Cruchet, P. H., & Pascal, H. (1988). *Correlation between ciguateric fish and damage to reefs in the Gambier Islands (French Polynesia)*. In Proceedings of the 6th International coral reef symposium, vol. 2, Australia, pp. 195–200.

Balech, E. (1995). *The genus Alexandrium Halim (Dinoflagellata)* (151 pp). Sherkin Island: Sherkin Island Marine Station.

Banner, A. H. (1976). Ciguatera: A disease from coral reefs. In N. A. Jones & R. Endean (Eds.), *Biology and geology of coral reefs* (Vol. 3, pp. 177–213). New York: Academic Press.

Barnes, M. K., Tilstone, G. H., Smyth, T. J., Widdicombe, C. E., Gloël, J., Robinson, C., & Suggett, D. J. (2015). Drivers and effects of Karenia mikimotoi blooms in the western English Channel. *Progress in Oceanography, 137*, 456–469.

Bauman, A. G., Burt, J. A., Feary, D. A., Marquis, E., & Usseglio, P. (2010). Tropical harmful algal blooms: An emerging threat to coral reef communities? *Marine Pollution Bulletin, 60*, 2117–2122.

Belin, C. (1991). Distribution of *Dinophysis* spp. and *Alexandrium minutum* along French coast since 1984 and their DSP and PSP toxicity levels. In T. J. Smayda & Y. Shimizu (Eds.), *Toxic phytoplankton blooms in the sea* (pp. 469–474). New York: Elsevier.

Bendis, B. J., Pigg, R. J., & Millie, D. F. (2004). Primary productivity of Florida red tides. In K. A. Steidinger, J. H. Landsberg, C. R. Tomas, & G. A. Vargo (Eds.), *Harmful algae 2002* (pp. 35–37). St. Petersburg: I.O.C of UNESCO.

Berdalet, E., Fleming, L. E., Gowen, R., Davidson, K., Hess, P., Backer, L. C., Moore, S. K., Hoagland, P., & Enevoldsen, H. (2016). Marine harmful algal blooms, human health and wellbeing: Challenges and opportunities in the 21st century. *Journal of the Marine Biological Association of the UK, 96*(1), 61–91.

Bhimachar, B. S., & George, P. C. (1950). Abrupt set-backs in the fisheries of the Malabar and Kanara coasts and 'Red water' phenomenon as their probable cause. *Proceedings of the Indian Academy of Science B, 31*, 339–350.

Borkman, D. G., Smayda, T. J., Tomas, C. R., York, R., Strangman, W., & Wright, J. L. (2012). Toxic *Alexandrium peruvianum* (Balech and de Mendiola) Balech and Tangen in Narragansett Bay, Rhode Island (USA). *Harmful Algae, 19*, 92–100.

Brand, L. E., Campbell, L., & Bresnan, E. (2012). Karenia: The biology and ecology of a toxic genus. *Harmful Algae, 14*, 156–178.

Brandenburg, K. M., Domis, L. N. S., Wohlrab, S., Krock, B., John, U., Scheppingen, Y., Donk, E., & Waal, D. B. V. (2017). Combined physical, chemical and biological factors shape *Alexandrium ostenfeldii* blooms in The Netherlands. *Harmful Algae, 63*, 146–153.

Bresnan, E., Fryer, R. J., Fraser, S., Smith, N., Stobo, L., Brown, N., & Turrell, E. (2017). The relationship between *Pseudo-nitzschia* (Peragallo) and domoic acid in Scottish shellfish. *Harmful Algae, 63*, 193–202.

Capone, D. G., Zehr, J. P., Paerl, H. W., Bergman, B., & Carpenter, E. J. (1997). *Trichodesmium*, a globally significant marine bacteria. *Science, 276*, 1221–1229.

Capone, D. G., Subramaniam, A., Montoya, J. P., Voss, M., Humborg, C., Johansen, A. M., Siefert, R. L., & Carpenter, E. J. (1998). An extensive bloom of the N2-fixing cyanobacterium, *Trichodesmium erythraeum* in the central Arabian Sea. *Marine Ecology Progress Series, 172*, 281–292.

Carpenter, E. J., Subramaniam, A., & Capone, D. G. (2004). Biomass and primary productivity of the cyanobacterium *Trichodesmium* spp. in the tropical N Atlantic Ocean. *Deep Sea Research Part A: Oceanographic Research Papers, 51*(2), 173–203.

Cembella, A. D., Quilliam, M. A., Lewis, N. I., Bauder, A. G., Dell'Aversano, C., Thomas, K., Jellett, J., & Cusack, R. R. (2002). The toxigenic marine dinoflagellate *Alexandrium tamarense* as the probable cause of mortality of caged salmon in Nova Scotia. *Harmful Algae, 1*, 313–325.

Chang, F. H. (2000). Pink blooms in the springs in Wellington harbour. *Aquaculture Update, 24*, 10–12.

Chinain, M., Germain, M., Deparis, X., Pauillac, S., & Legrand, A. M. (1999). Seasonal abundance and toxicity of the dinoflagellates *Gambierdiscus* spp. (Dinophyceae), the causative agent of ciguatera in Tahiti, French Polynesia. *Marine Biology, 135*, 259–267.

Cho, E., & Costas, E. (2004). Rapid monitoring for the potentially ichthyotoxic dinoflagellate *Cochlodinium polykrikoides* in Korean coastal waters using fluorescent probe tools. *Journal of Plankton Research, 26*, 175–180.

Cloern, J. E. (2001). Our evolving conceptual model of the coastal eutrophication problem. *Marine Ecology Progress Series, 210*, 223–253.

Curtiss, C. C., Langlois, G. W., Busse, L. B., Mazzillo, F., & Silver, M. W. (2008). The emergence of Cochlodinium along the California coast (USA). *Harmful Algae, 7*, 337–346.

D'Silva, M. S., Anil, A. C., Naik, R. K., & D'Costa, P. M. (2012). Algal blooms: A perspective from the coasts of India. *Natural Hazards, 63*, 1225–1253.

Dabrowski, T., Lyons, K., Nolan, G., Berry, A., Cusack, C., & Silke, J. (2016). Harmful algal bloom forecast system for SW Ireland. Part I: Description and validation of an operational forecasting model. *Harmful algae, 53*, 64–76.

Dahl, E., & Tangen, K. (1993). 25 years experience with Gyrodinium aureolum in Norwegian waters. In *Developments in marine biology.*

Dale, B., & Yentsch, C. M. (1978). Red tide and paralytic shellfish poisoning. *Oceanus, 21*, 41–49.

Dale, B., Yentsch, C. M., & Hurst, J. W. (1978). Toxicity in resting cyst of the red tide Gonyaulax excavata from deeper water coastal sediments. *Science, 201*, 1223–1225.

Davidson, K., Miller, P., Wilding, T. A., Shutler, J., Bresnan, E., Kennington, K., & Swan, S. (2009). A large and prolonged bloom of *Karenia mikimotoi* in Scottish waters in 2006. *Harmful Algae, 8*(2), 349–361.

Davidson, K., Anderson, D. M., Mateus, M., Reguera, B., Silke, J., Sourisseau, M., & Maguire, J. (2016). Forecasting the risk of harmful algal blooms. *Harmful Algae, 53*, 1–7.

Dawson, E., Aleem, A., & Halstead, B. (1955). *Marine algae from Palmyra Islands with special reference to the feeding habits and toxicology of reef fishes*. Allan Hancock Publications of the University of Southern California. Occasional paper 17.

Deeds, J. R., Wiles, K., Heideman, V. I. G. B., White, K. D., & Abraham, A. (2010). First U.S. reports of shellfish harvesting closures due to confirmed okadaic acid in Texas Gulf coast oysters. *Toxicon, 55*, 1138–1146.

Dhar, K., Sinha, A., Gaur, P., Goel, R., Chopra, V. S., & Bajaj, U. (2015). Pattern of adverse drug reactions to antibiotics commonly prescribed in department of medicine and pediatrics in a tertiary care teaching hospital, Ghaziabad. *Journal of Applied Pharmaceutical Science, 5*(4), 78–82.

Diaz, R. J., & Rosenberg, R. (2008). Spreading dead zones and consequences for marine ecosystems. *Science, 321*(5891), 926–929.

Diaz, P. A., Molinet, C., Seguel, B., Manuel Díaz, M., Labra, G., & Figueroa, R. I. (2014). Coupling planktonic and benthic shifts during a bloom of *Alexandrium catenella* in southern Chile: Implications for bloom dynamics and recurrence. *Harmful Algae, 40*, 9–22.

Díaz, P. A., Ruiz-Villarreal, M., Pazos, Y., Moita, T., & Reguera, B. (2016). Climate variability and Dinophysis acuta blooms in an upwelling system. *Harmful Algae, 53*, 145–159.

Dodge, J. D. (1982). *Marine dinoflagellates of the British Isles* (2nd ed.303 pp). London: HM Stat Office.

Driggers, W. B., III, Campbell, M. D., Debose, A. J., Hannan, K. M., Hendon, M. D., Martin, T. L., & Nichols, C. C. (2016). Environmental conditions and catch rates of predatory fishes associated with a mass mortality on the West Florida Shelf. *Estuarine and Coastal Shelf Science, 168*, 40–49.

Edwards, M., & Richardson, A. J. (2004). Impact of climate change on marine pelagic phenology and trophic mismatch. *Nature, 430*, 881–884.

Evans, D. (1975). *The occurrence of Gyrodinium aureolum in the eastern Irish Sea, 1975*. Lancashire and Western Irish Sea Fisheries Joint Committee scientific report, pp. 85–89.

FAO. (2004). *The state of world fisheries and aquaculture (SOFIA) 2004*. Rome: Food and Agriculture Organization.

Fauquier, D. A., Flewelling, L. J., Maucher, J. M., Keller, M., Kinsel, M. J., Johnson, C. K., Henry, M., Gannon, J. G., Ramsdell, J. S., & Landsberg, J. H. (2013). Brevetoxicosis in seabirds naturally exposed to *Karenia brevis* blooms along the central west coast of Florida. *Journal of Wildlife Diseases, 49*(2), 246–260.

Faust, M. A., & Gulledge, R. A. (2002). Identifying harmful marine dinoflagellates. *Contributions from the United States National Herbarium, 42*, 1–144.

Figueroa, R. I., Bravo, I., & Garcés, E. (2006). Multiples routes of sexuality in *Alexandrium taylori* (Dinophyceae) in culture. *Journal of Phycology, 42*, 1028–1039.

Figueroa, R. I., Bravo, I., & Garcés, E. (2008). The significance of sexual versus asexual cyst formation in the life cycle of the noxious dinoflagellate *Alexandrium peruvianum*. *Harmful Algae, 7*, 653–663.

Fire, S. E., Wang, Z., Byrd, M., Whitehead, H. R., Paternoster, J., & Morton, S. L. (2011). Co-occurrence of multiple classes of harmful algal toxins in bottlenose dolphins (*Tursiops truncatus*) stranding during an unusual mortality event in Texas, USA. *Harmful Algae, 10*(3), 330–336.

Fleming, L., Baden, E., Bean, J. A., Weiman, R., & Blythe, D. G. (1998). Seafood toxin diseases: Issues in epidemiology and community outreach. In B. Reguera, J. Blanco, M. L. Fernandez, & T. Wyatt (Eds.), *Harmful Algae* (pp. 245–248). Galicia: Xunta de Galicia and Intergovernmental Oceanographic Commission of UNESCO.

Fleming, L. E., Backer, L. C., & Baden, D. G. (2005). Overview of aerosolized Florida red tide toxins: Exposures and effects. *Environmental Health Perspectives, 113*, 618–620.

Fleming, L. E., Kirkpatrick, B., Backer, L. C., Walsh, C. J., Nierenberg, K., Clark, J., Reich, A., Hollenbeck, J., Benson, J., Cheng, Y. S., Naar, J., Pierce, R., Bourdelais, A. J., Abraham, W. M., Kirkpatrick, G., Zaias, J., Wanner, A., Mendes, E., Shalat, S., Hoagland, P., Stephan, W., Bean, J., Watkins, S., Clarke, T., Byrne, M., & Baden, D. G. (2011). Review of Florida red tide and human health effects. *Harmful Algae, 20*, 224–233.

Fraga, S., Marino, J., Bravo, I., Miranda, A., Campos, M.J., Sanchez, F.J., Costas, E., Cabanas, J.M., & Blanco, J. (1984). *Red tides and shellfish poisoning in Galicia (NW Spain)*. In: ICES Special meeting on the causes, dynamics and effects of exceptional marine blooms and related events, Copenhagen, 4–5 October 1984, p. 5.

Fraga, S., Anderson, D. M., Bravo, I., Reguera, B., Steidinger, K. A., & Yentsch, C. M. (1988). Influence of upwelling relaxation on dinoflagellates and shellfish toxicity in Ria de Vigo, Spain. *Estuarine, Coastal and Shelf Science, 27*(4), 349–361.

Fromentin, J. M., & Planque, B. (1996). *Calanus* and environment in the eastern North Atlantic, 2. Influence of the North Atlantic Oscillation on *Calanus finmarchicus* and *C. helgolandicus*. *Marine Ecology Progress Series, 134*, 111–118.

Fukuyo, Y., Takano, H., Chihara, M., & Matsuoka, K. (1990). *Red tide organisms in Japan. An illustrated taxonomic guide* (430 pp). Tokyo: Uchida Rokakuho Publishing Press.

Furey, A., Braña-Magdalena, A., Lehane, M., Moroney, C., James, K. J., Satake, M., & Yasumoto, T. (2002). Determination of azaspiracids in shellfish using liquid chromatography–tandem electrospray mass spectrometry. *Rapid Communications in Mass Spectrometry, 16*, 238–242.

Furey, A., Moroney, C., Magdalena, A. B., Saez, M. J. F., Lehane, M., & James, K. J. (2003). Geographical, temporal, and species variation of the polyether toxins, azaspiracids, in shellfish. *Environmental Science and Technology, 37*, 3078–3084.

Garcés, E., Masó, M., & Camp, J. (1999). A recurrent and localized dinoflagellate bloom in a Mediterranean beach. *Journal of Plankton Research, 21*(12), 2373–2391.

Garcia, V. M. T., & Purdie, D. A. (1994). Primary production studies during a Gyrodinium cf. aureolum (Dinophyceae) bloom in the western English Channel. *Marine Biology, 119*(2), 297–305.

García-Soto, C., & Pingree, R. D. (2009). Spring and summer blooms of phytoplankton(SeaWiFS/MODIS) along a ferry line in the Bay of Biscay and western English Channel. *Continental Shelf Research, 29*, 1111–1122.

Gentien, P. (1998). Bloom dynamics and ecophysiology of the *Gymnodinium mikimotoi* species complex. In D. L. Anderson, A. D. Cembella, & G. M. Hallegraeff (Eds.), *Physiological ecology of harmful algal blooms, NATO ASI Series* (pp. 155–173). Berlin: Springer.

Gentien, P., Lunven, M., Lazure, P., Youenou, A., & Crassous, M. P. (2007). Motility and autotoxicity in *Karenia mikimotoi* (Dinophyceae). *Philosophical Transaction of the Royal Society of London. Series B, Biological Sciences, 362*(1487), 1937–1946.

Gessner, B. D., Bell, P., Doucette, G. J., Moczydlowski, E., Poli, M. A., Van Dolah, F., & Hall, S. (1997a). Hypertension and identification of toxin in human urine and serum following a cluster of mussel-associated paralytic shellfish poisoning outbreaks. *Toxicon, 35*(5), 711–722.

Gessner, B. D., Middaugh, J. P., & Doucette, G. J. (1997b). Paralytic shellfish poisoning in Kodiak, Alaska. *Western Journal of Medicine, 167*(5), 351–353.

Gestal-Otero, J. J. (2000). Nonneurotoxic toxins. In L. M. Botana (Ed.), *Seafood and fresh water toxins* (pp. 45–64). Boca Raton: CRC press.

Gran, H. H. (1897). Botanik. Prophyta: Diatomaceae, Silicoflagellata og Cilioflagellata. *Den Norske Nordhavs Expedition, 1876-1878*(7), 1–36.

Gran, H. H., & Angst, E. C. (1931). *Plankton diatoms of puget sound* (Vol. 7). Seattle: University Press.

Graneli, E., & Turner, J. T. (Eds.). (2008). *Ecology of harmful algae*. Berlin: Springer.

Grattan, L. M., Holobaugh, S., & Morris, J. G., Jr. (2016). Harmful algal blooms and public health. *Harmful Algae, 57*, 2–8.

Griffith, A. W., & Gobler, C. J. (2016). Temperature controls the toxicity of the ichthyotoxic dinoflagellate *Cochlodinium polykrikoides*. *Marine Ecology Progress Series, 545*, 63–76.

Guinder, V. A., Carcedo, M. C., Buzzi, N., Molinero, J. C., Abbate, C. L., Melisa, F. S., & Kühn, S. (2017). Ephemeral parasitism on blooming diatoms in a temperate estuary. *Marine and Freshwater Research, 69*(1), 128–133.

Guiry, M. D., & Guiry, G. M. (2011). *AlgaeBase*. Galway: World-wide Electronic Publication, National University of Ireland. http://www.algaebase.org.

Habekost, R., Fraser, I., & Halstead, B. (1955). Toxicology: Observations on toxic marine algae. *Journal of the Washington Academy of Sciences, 45*, 101–103.

Hakanen, P., Suikkanen, S., Franzén, J., Franzén, H., Kankaanpää, H., & Kremp, A. (2012). Bloom and toxin dynamics of *Alexandrium ostenfeldii* in a shallow embayment at the SW coast of Finland, northern Baltic Sea. *Harmful Algae, 15*, 91–99.

Hallegraeff, G. M. (1991). *Aquaculturists guide to harmful Australian microalgae* (1st ed.111 pp). Hobart: Fish. Ind. Train. Board, Tasmania/CSIRO Div. Fisher.

Hallegraeff, G. M. (1993). A review of harmful algal blooms and their apparent global increase. *Phycologia, 32*, 79–99.

Hallegraeff, G. M. (2010). Ocean climate change, phytoplankton community responses, and harmful algal blooms: A formidable predictive challenge. *Journal of Phycology, 46*, 220–235.

Hallegraeff, G., Marshall, J., Valentine, J., & Hardiman, S. (1998). Short cyst-dormancy period of an Australian isolate of the toxic dinoflagellate *Alexandrium catenella*. *Marine and Freshwater Research, 49*, 415–420.

Halstead, B. W., & Schantz, E. Z. (1984). *Paralytic shellfish poisoning*. Geneva: World Health Organization.

Harrison, P. J., Furuya, K., Glibert, P. M., Xu, J., Liu, H. B., Yin, K., Lee, J. H. W., Anderson, D. M., Gowen, R., Al-Azri, A. R., & Ho, A. Y. T. (2011). Geographical distribution of red and green Noctiluca scintillans. *Chinese Journal of Oceanology and Limnology, 29*(4), 80–831.

Hashimoto, Y., Kamiy, H., Yamazato, K., & Nozawa, K. (1976). Occurrence of toxic blue green alga inducing skin dermatitis in Okinawa. In A. Ohsaka, K. Hayashi, & Y. Sawai (Eds.), *Animal, Plant and microbial toxins* (pp. 333–338). New York: Plenum.

Hausmann, K., Hulsmann, N., & Radek, R. (2003). Protistology (3rd ed.). In *E. Scheizerbart'sche Verlagsbuchhandlung*.

He, R., McGillicuddy, D. J., Keafer, B. A., & Anderson, D. M. (2008). Historic 2005 toxic bloom of Alexandrium fundyense in the western Gulf of Maine: 2. Coupled biophysical numerical modeling. *Journal of Geophysical Research: Oceans, 113*, C7.

Head, M. J. (1996). Modern dinoflagellate cysts and their biological affinities. In J. Jansonius & D. C. McGregor (Eds.), *Palynology, principles and applications* (Vol. 3, pp. 1197–1248). College Station: American Association of Stratigraphic Palynologists Foundation.

Heisler, J., Glibert, P., Burkholder, J., Anderson, D., Cochlan, W., Dennison, W., Gobler, C., Dortch, Q., Heil, C., Humphries, E., Lewitus, A., Magnien, R., Marshall, H., Sellner, K., Stockwell, D., Stoecker, D., & Suddleson, M. (2008). Eutrophication and harmful algal blooms: A scientific consensus. *Harmful Algae, 8*, 3–13.

Hernández-Becerril, D. U., Alonso-Rodríguez, R., Álvarez-Góngora, C., Barón-Campis, S. A., Ceballos-Corona, G., Herrera-Silveira, J., Meave Del Castillo, M. E., Juárez-Ruíz, N., Merino-Virgilio, F., Morales-Blake, A., Ochoa, J. L., Orellana-Cepeda, E., Ramírez-Camarena, C., & Rodríguez-Salvador, R. (2007). Toxic and harmful marine phytoplankton and microalgae (HABs) in Mexican coasts. *Journal of Environmental Science and Health Part A. Toxic/ Hazardous Substances Environmental Engineering, 42*(10), 1349–1363.

Hinder, S., Hays, G., Brooks, C., Davies, A., Edwards, M., Walne, A., & Gravenor, M. (2011). Toxic marine microalgae and shellfish poisoning in the British Isles: History, review of epidemiology, and future implications. *Environmental Health, 10*(1), 54.

Hisem, D., Hrouzek, P., Tomek, P., Tomšíčková, J., Zapomělová, E., Skácelová, K., & Kopecký, J. (2011). Cyanobacterial cytotoxicity versus toxicity to brine shrimp Artemia salina. *Toxicon, 57*(1), 76–83.

Hitchcock, G. L., Kirkpatrick, G., Minnett, P., & Palubok, V. (2010). Net community production and dark community respiration in a *Karenia brevis* (Davis) bloom in West Florida coastal waters, USA. *Harmful Algae, 9*, 351–358.

HKEPD. (2014). *Monitoring of solid waste in Hong Kong – Waste statistics for 2012*. Hong Kong SAR: Environmental Protection Department.

Hoagland, P., Jin, D., Beet, A., Kirkpatrick, B., Reich, A., Ullmann, S., Fleming, L. E., & Kirkpatrick, G. (2014). The human health effects of Florida Red Tide (FRT) blooms: An expanded analysis. *Environment International, 68*, 144–153.

Holligan, P. M., Williams, P. J. L., Purdie, D., & Harris, R. P. (1984). Photosynthesis, respiration and nitrogen supply of plankton populations in stratified, frontaland tidally mixed shelf waters. *Marine Ecology Progress Series, 17*, 201–213.

Howard, M. D. A., Jones, A. C., Schnetzer, A., Countway, P. D., Tomas, C. R., Kudela, R. M., Hayashi, K., Chia, P., & Caron, D. A. (2012). Quantitative real-time PCR for *Cochlodinium fulvescens* (Dinophyceae), a potentially harmful dinoflagellate from California coastal waters. *Journal of Phycology, 48*, 384–393.

Huang, C., & Qi, H. (1997). The abundance cycle and influence factors on red tide phenomena of Noctiluca scintillans (Dinophyceae) in Dapeng Bay, the South China Sea. *Journal of Plankton Research, 19*(3), 303–318. https://doi.org/10.1093/plankt/19.3.303.

Ishida, H., Muramatsu, N., Nukaya, H., Kosuge, T., & Tsuji, K. (1996). Study on neurotoxic shellfish poisoning involving the oyster, *Crassostrea gigas*, in New Zealand. *Toxicon, 34*(9), 1050–1053.

Islam, M. S., & Tanaka, M. (2004). Impacts of pollution on coastal and marine ecosystems including coastal and marine fisheries and approach for management: A review and synthesis. *Marine Pollution Bulletin, 48*, 624–649.

Iwataki, M., Kawami, H., & Matsuoka, K. (2007). *Cochlodinium fulvescens* sp. nov. (Gymnodiniales, Dinophyceae), a new chain-forming unarmored dinoflagellate from Asian coasts. *Phycological Research, 55*, 231–239.

Iwataki, M., Hansen, G., Moestrup, O., & Matsuoka, K. (2010). Ultrastructure of the harmful unarmored dinoflagellate *Cochlodinium polykrikoides* (Dinophyceae) with reference to the apical groove and flagellar apparatus. *The Journal of Eukaryotic Microbiology, 57*, 308–321.

James, K.J., Furey, A., Satake, M., & Yasumoto, T. (2000). *Azaspiracid poisoning (AZP): A new shell fish toxic syndrome in Europe*. In Ninth International conference on harmful algal blooms, pp. 250–253.

James, K. J., Lehane, M., Moroney, C., Fernandez-Puente, P., Satake, M., Yasumoto, T., & Furey, A. (2002). Azaspiracid shellfish poisoning: Unusual toxin dynamics in shellfish and the increased risk of acute human in toxications. *Food Additives and Contaminants, 19*, 555–561.

James, K. J., Sáez, M. J. F., Furey, A., & Lehane, M. (2004). Azaspiracid poisoning, the food-borne illness associated with shellfish consumption: A review. *Food Additives and Contaminants, 21*, 879–892.

James, K. J., O'Driscoll, D., Fernandez, J. G., & Furey, A. (2008). Azaspiracids: Chemistry, bioconversion and determination. In L. M. Botana (Ed.), *Seafood and freshwater toxins: Pharmacology, physiology and detection* (pp. 763–773). New York: CRC Press, Taylor and Francis.

James, K. J., Carey, B., O'Halloran, J., Van Pelt, F. N. A. M., & Ŝkrabákovẚ, Z. (2010). Shellfish toxicity: Human health implications of marine algal toxins. *Epidemiology and Infection, 138*(07), 927–940.

Katsuhide, M. N., Goh, Y., Yuichiro, N., Yoshinaga, Y. I., & Hikawa, A. (2017). Germination fluctuation of toxic *Alexandrium fundyense* and *A. pacificum* cysts and the relationship with bloom occurrences in Kesennuma Bay, Japan. *Harmful Algae, 62*, 52–59.

Kawabata, T. (1989). Regulatory aspects of marine biotoxins in Japan. In S. Natori, K. Hashimoto, & Y. Ueno (Eds.), *Mycotoxins and phycotoxins* (pp. 469–476). Amsterdam: Elsevier.

Keafer, B. A., Churchill, J. H., McGillicuddy, D. J., Jr., & Anderson, D. M. (2005). Bloom development and transport of toxic *Alexandrium fundyense* populations within a coastal plume in the Gulf of Maine. *Deep Sea Research Part II, 52*, 2674–2697.

Kerbrat, A. S., Amzil, Z., Pawlowiez, R., Golubic, S., Sibat, M., Darius, H. T., Chinain, M., Kerbrat, A. S., Darius, H. T., Pauillac, S., Chinain, M., & Laurent, D. (2010). Detection of ciguatoxin-like and paralysing toxins in *Trichodesmium* spp. from New Caledonia lagoon. *Marine Pollution Bulletin, 61*, 360–366.

Kerbrat, A. S., Amzil, Z., Pawlowiez, R., Golubic, S., Sibat, M., Darius, H. T., et al. (2011). First evidence of palytoxin and 42-hydroxy-palytoxin in the marine cyanobacterium Trichodesmium. *Marine Drugs, 9*(4), 543–560.

Kim, H. G. (1997). Recent harmful algal blooms and mitigation strategies in Korea. *Ocean Res. (Seoul), 19*, 185–192.

Kim, H. G. (2006). Mitigation and controls of HABs. In E. Granéli & J. T. Turner (Eds.), *Ecology of harmful algae* (pp. 327–338). Berlin/Heidelberg: Springer.

Kim, D., Oda, T., Muramatsu, T., Kim, D., Matsuyama, Y., & Honjo, T. (2002). Possible factors responsible for the toxicity of *Cochlodinium polykrikoides*, a red tide phytoplankton. *Comparative Biochemistry and Physiology Part C Toxicology and Pharmacology, 132*, 415–423.

Kim, D. W., Jo, Y. H., Choi, J. K., Choi, J. G., & Bi, H. (2016). Physical processes leading to the development of an anomalously large *Cochlodinium polykrikoides* bloom in the East Sea/Japan Sea. *Harmful Algae, 55*, 250–258.

Kim, J. H., Tillmann, U., Adams, N. G., Krock, B., Stutts, W. L., Deeds, J. R., et al. (2017). Identification of Azadinium species and a new azaspiracid from *Azadinium poporum* in Puget sound, Washington state, USA. *Harmful Algae, 68*, 152–167.

Kirkpatrick, B., Fleming, L. E., Bean, J. A., Nierenberg, K., Backer, L. C., Cheng, Y. S., Pierce, R., Reich, A., Naar, J., Wanner, A., Abraham, W. M., Zhou, Y., Hollenbeck, J., & Baden, D. G. (2011). Aerosolized red tide toxins (brevetoxins) and asthma: Continued health effects after 1h beach exposure. *Harmful Algae, 10*, 138–143.

Knaack, J. S., Porter, K. A., Jacob, J. T., Sullivan, K., Forester, M., Wang, R. Y., & Thomas, J. (2016). Case diagnosis and characterization of suspected paralytic shellfish poisoning in Alaska. *Harmful Algae, 57*, 45–50.

Kofoid, C. A., & Swezy, O. (1921). The free-living unarmored dinoflagellata. *Memoirs of the Univerisity California, 5*, 1–562.

Kohler, S. T., & Kohler, C. C. (1992). Dead bleached coral provides new surfaces for dinoflagellates implicated in ciguatera fish poisonings. *Environmental Biology of Fishes, 35*, 413–416.
Kudela, R. M., & Gobler, C. J. (2012). Harmful dinoflagellate blooms caused by *Cochlodinium* sp.: Global expansion and ecological strategies facilitating bloom formation. *Harmful Algae, 14*, 71–86.
Kuwata, A., & Jewson, D. H. (2015). In S. Ohtsuka, T. Suzaki, T. Horiguchi, N. Suzuki, F. Not (Eds.), *Ecology and evolution of marine diatoms and parmales* (pp. 251–275). Tokyo: Springer.
Kvitek, R. G., Goldberg, J. D., Smith, G. J., & Silver, M. W. (2008). Domoic acid contamination within eight representative species from the benthic food web of Monterey Bay, California, USA. *Marine Ecology Progress Series, 367*, 35–47.
Lagos, N. (2003). Paralytic shellfish poisoning phycotoxins: Occurrence in South America. *Comments on Toxicology, 9*, 175–193.
Landsberg, J. H. (2002). The effects of harmful algal blooms on aquatic organisms. *Reviews of Fisheries Science, 10*, 113–390.
Laurent, D., Kerbrat, A., Darius, H. T., Girard, E., Golubic, S., Benoit, E., Sauviat, M. P., Chinain, M., Molgo, J., & Pauillac, S. (2008). Are cyanobacteria involved in Ciguatera fish poisoning-like outbreaks in New Caledonia? *Harmful Algae, 7*, 827–838.
Lebour, M. V. (1925). *The Dinoflagellates of Northern Seas* (250 pp). Plymouth: Marine Biology Association.
Lee, M. O., & Choi, J. H. (2009). Distributions of water temperature and salinity in the Korea Southern coastal water during *Cochlodinium polykrikoides* Blooms. *Journal of the Korean Society for Marine Environment & Energy, 12*(4), 235–247.
Lee, Y. J., Choi, J. K., Kim, E. K., Youn, S. H., & Yang, E. J. (2008). Field experiments on mitigation of harmful algal blooms using a Sophorolipid—Yellow clay mixture and effects on marine plankton. *Harmful Algae, 7*(2), 154–162.
Lewis, R. J. (2006). Ciguatera: Australian perspectives on a global problem. *Toxicon, 48*, 799–809.
Lewitus, A. J., Horner, R. A., Caron, D. A., Garcia-Mendoza, E., Hickey, B. M., Hunter, M., & Lessard, E. J. (2012). Harmful algal blooms along the North American west coast region: History, trends, causes, and impacts. *Harmful Algae, 19*, 133–159.
Lilly, E. L., Halanych, K. M., & Anderson, D. M. (2007). Species boundaries and global biogeography of the *Alexandrium tamarense* complex (Dinophyceae)1. *Journal of Phycology, 43*(6), 1329–1338. https://doi.org/10.1111/j.1529-8817.2007.00420.x.
Lim, A. S., Jeong, H. J., Kim, J. H., & Lee, S. Y. (2017). Control of ichthyotoxic *Cochlodinium polykrikoides* using the mixotrophic dinoflagellate *Alexandrium pohangense:* A potential effective sustainable method. *Harmful Algae, 63*, 109–118.
Lirdwitayaprasit, T., Nishio, S., Montani, S., & Okaichi, T. (1990). The biochemical processes during cyst formation in *Alexandrium catenella*. In E. Gráneli, B. Sund-Strom, L. Edler, & D. M. Anderson (Eds.), *Toxic marine phytoplankton* (pp. 294–299). New York: Elsevier.
Liu, X. J., & Wong, C. K. (2006). Seasonal and spatial dynamics of *Noctiluca scintillans* in a semi-enclosed bay in the northeastern part of Hong Kong. *Botanica Marina, 49*, 145–150.
Liu, G., Fan, C., Zhong, J., Zhang, L., Ding, S., Yan, S., & Han, S. (2010). Using hexadecyl trimethyl ammonium bromide (CTAB) modified clays to clean the Microcystis aeruginosa blooms in Lake Taihu, China. *Harmful Algae, 9*(4), 413–418.
Liu, R., Men, C., Liu, Y., Yu, W., Xu, F., & Shen, Z. (2016a). Spatial distribution and pollution evaluation of heavy metals in Yangtze Estuary sediment. *Marine Pollution Bulletin, 110*(1), 564–571.
Liu, Y., Cao, X., Yu, Z., Song, X., & Qiu, L. (2016b). Controlling harmful algae blooms using aluminum-modified clay. *Marine Pollution Bulletin, 103*(1–2), 211–219.
Ljubešić, Z., Bosak, S., Viličić, D., Borojević, K. K., Marić, D., Godrijan, J., & Dakovac, T. (2011). Ecology and taxonomy of potentially toxic *Pseudo-nitzschia* species in Lim Bay (north-eastern Adriatic Sea). *Harmful Algae, 10*(6), 713–722.
Lu, G., Song, X., Yu, Z., Cao, X., & Yuan, Y. (2015). Environmental effects of modified clay flocculation on Alexandrium tamarense and paralytic shellfish poisoning toxins (PSTs). *Chemosphere, 127*, 188–194.

Lu, X., Wang, Z., Guo, X., Gu, Y., Liang, W., & Liu, L. (2017). Impacts of metal contamination and eutrophication on dinoflagellate cyst assemblages along the Guangdong coast of southern China. *Marine Pollution Bulletin, 120*(1–2), 239–249.

Ma, J., & Liu, W. (2002). Effectiveness and mechanism of potassium ferrate (VI) preoxidation for algae removal by coagulation. *Water Research, 36*(4), 871–878.

Magdalena, A. B., Lehane, M., Krys, S., Fernandez, M. L., Furey, A., & James, K. J. (2003a). The first identification of azaspiracids in shellfish from France and Spain. *Toxicon, 42*, 105–108.

Magdalena, A. B., Lehane, M., Moroney, C., Furey, A., & James, K. J. (2003b). Food safety implications of the distribution of azaspiracids in the tissue compartments of scallops (*Pecten maximus*). *Food Additives and Contaminants, 20*, 154–160.

Mahoney, J. B., Olsen, P., & Cohn, M. (1990). Blooms of a dinoflagellate Gyrodinium cf. aureolum in New Jersey coastal waters and their occurrence and effects worldwide. *Journal of Coastal Research, 6*(1), 121–135.

Maldonado, M. T. (2002). The effect of Fe and Cu on growth and domoic acid production by *Pseudo-nitzschia* multiseries and *Pseudo-nitzschia* australis. *Limnology and Oceanography, 47*, 515–526. https://doi.org/10.4319/lo.2002.47.2.0515.

Margalef, R. (1961). Hidrografia y fitoplancton de un area marina de la costa meridional de Puerto Rico. *Investigación Pesquera, 18*, 76–78.

Maruyama, T., Yamada, R., Usui, K., Suzuki, H., & Yoshida, T. (1987). Removal of marine red tide planktons with acid treated clay. *Bulletin of the Japanese Society of Scientific Fisheries, 53*(10), 1811–1819.

Matrai, P., Thompson, B., & Keller, M. (2005). *Alexandrium* spp. from eastern Gulf of Maine populations: Circannual excystment. *Deep Sea Research Part II, 52*, 2560–2568.

Matsuoka, K., Iwataki, M., & Kawami, H. (2008). Morphology and taxonomy of chainforming species of the genus *Cochlodinium* (Dinophyceae). *Harmful Algae, 7*, 261–270.

Matsuoka, K., Takano, Y., Kamrani, E., Rezai, H., Puthiyedathu, S. T., Al-Gheilani, H. M., & Zhenqing, M. (2010). Study on *Cochlodinium polykrikoides* Margalef (Gymnodiniales, Dinophyceae) in the Oman Sea and the Persian Gulf from august 2008 to august 2009. *Current Development in Oceanography, 1*(3), 153–171.

Matsuoka, K., Ikeda, Y., Kaga, S., Kaga, Y., & Ogata, T. (2018). Repercussions of the great East Japan earthquake tsunami on ellipsoidal *Alexandrium* cysts (Dinophyceae) in Ofunato Bay, Japan. *Marine Environmental Research, 135*, 123–135.

McCarron, P., Kilcoyne, J., & Hess, P. (2008). Effects of cooking and heat treatment on concentration and tissue distribution of okadaic acid and dinophysistoxin-2 in mussels (*Mytilus edulis*). *Toxicon, 51*, 1081–1089.

McDonald, S. M., Sarno, D., & Zingone, A. (2007). Identifying Pseudo-nitzschia species in natural samples using genus-specific PCR primers and clone libraries. *Harmful Algae, 6*, 849–860.

McGillicuddy, D., Signell, R., Stock, C., Keafer, B., Keller, M., Hetland, R., & Anderson, D. (2003). A mechanism for offshore initiation of harmful algal blooms in the coastal Gulf of Maine. *Journal of Plankton Research, 25*(9), 1131–1138.

McGillicuddy, D., Anderson, D. M., Lynch, D. R., & Townsend, D. W. (2005). Mechanisms regulating large-scale seasonal fluctuations in *Alexandrium fundyense* populations in the Gulf of Maine: Results from a physical–biological model. *Deep Sea Research Part II, 52*(19–21), 2698–2714.

McLaughlin, J. B., Fearey, D. A., Esposito, T. A., & Porter, K. A. (2011). Paralytic shellfish poisoning – Southeast Alaska, May–June 2011. *Morbidity and Mortality Weekly Report, 60*, 1554–1556.

Méndez, S. M., Ferrario, M., & Cefarelli, A. O. (2012). Description of toxigenic species of the genus *Pseudo-nitzschia* in coastal waters of Uruguay: Morphology and distribution. *Harmful Algae, 19*, 53–60.

Miao, C. G., Tang, Y., Zhang, H., Wu, Z. Y., & Wang, X. Q. (2014). Harmful algae blooms removal from fresh water with modified vermiculite. *Environmental Technology, 35*(3), 340–346 (in Chinese).

Miller, P. E., & Scholin, C. A. (1996). Identification of cultured *Pseudo-nitzschia* (Bacillariophyceae) using species-specific LSU rRNA-targeted fluorescent probes. *Journal of Phycology, 32*, 646–655.

Miller, P. E., & Scholin, C. A. (1998). Identification and enumeration of cultured and wild *Pseudo-nitzschia* (Bacillariophyceae) using species-specific LSU rRNA-targeted fluorescent probes and filter-based whole cell hybridization. *Journal of Phycology, 34*, 371–382.

Miller, W. D., Harding, L. W., & Adolf, J. E. (2006). Hurricane Isabel generated an unusual fall bloom in Chesapeake Bay. *Geophysical Research Letters, 33*(6), L06612.

Miranda, L. N., Zhuang, Y., Zhang, H., & Lin, S. (2012). Phylogenetic analysis guided by intragenomic SSU rDNA polymorphism refines classification of "Alexandrium tamarense" species complex. *Harmful Algae, 16*, 35–48. https://doi.org/10.1016/j.hal.2012.01.002.

Mooney, B. D., Nichols, P. D., De Salas, M. F., & Hallegraeff, G. M. (2007). Lipid, fatty acid, and sterol composition of eight species of Kareniaceae (Dinophyta): Chemotaxonomy and putative lipid Phytotoxins. *Journal of Phycology, 43*(1), 101–111.

Morris, P. D., Campbell, D. S., Taylor, T. J., & Freeman, J. I. (1991). Clinical and epidemiological features of neurotoxic shellfish poisoning in North Carolina. *American Journal of Public Health, 81*(4), 471–474.

Murata, A., Nagashima, Y., & Taguchi, S. (2012). N: P ratios controlling the growth of the marine dinoflagellate Alexandrium tamarense: Content and composition of paralytic shellfish poison. *Harmful Algae, 20*, 11–18.

Narahashi, T., & Moore, J. W. (1968). Neuroactive agents and nerve membrane conductances. *The Jounal of General Physiology, 51*(5 pt. 2), 93.

Narayana, S., Chitra, J., Tapase, S. R., Thamke, V., Karthick, P., Ramesh, C., Murthy, K. N., Ramasamy, M., Kodam, K. M., & Mohanraju, R. (2014). Toxicity studies of *Trichodesmium erythraeum* (Ehrenberg, 1830) bloom extracts, from Phoenix Bay, Port Blair, Andamans. *Harmful Algae, 40*, 34–39.

National Research Council. (2006). *America's lab report: Investigations in high school science*. Washington, DC: National Academies Press.

Natsuike, M., Matsuno, K., Hirawake, T., Yamaguchi, A., Nishino, S., & Imai, I. (2017). Possible spreading of toxic *Alexandrium tamarense* blooms on the Chukchi Sea shelf with the inflow of Pacific summer water due to climatic warming. *Harmful Algae, 61*, 80–86.

Neely, T., & Campbell, L. (2006). A modified assay to determine hemolytic toxin variability among *Karenia* clones isolated from the Gulf of Mexico. *Harmful Algae, 5*(5), 592–598.

Neves, R. A., Contins, M., & Nascimento, S. M. (2018). Effects of the toxic benthic dinoflagellate Ostreopsis cf. ovata on fertilization and early development of the sea urchin Lytechinus variegatus. *Marine Environmental Research, 135*, 11–17.

O'Neil, J. M., Davis, T. W., Burford, M. A., & Gobler, C. J. (2012). The rise of harmful cyanobacteria blooms: The potential roles of eutrophication and climate change. *Harmful Algae, 14*, 313–334.

Okaichi, T. (1989). Red tide problems in the Seto Inland Sea, Japan. In T. Okaichi, D. M. Anderson, & T. Nemoto (Eds.), *Red tides: Biology, environmental science, and toxicology* (pp. 137–142). New York: Elsevier.

Okaichi, T., & Nishio, S. (1976). Identification of ammonia as the toxic principle of red tide of Noctiluca miliaris. *Bulletin of the Plankton Society of Japan, 23*, 75–80.

Okamura, K. (1916). Akashio ni Tsuite.([nl]On red tides). *Suisan Koushu Sikenjo Kenkyu Hokoku, 12*, 26–41.

Oshima, Y., Bolch, C., & Hallegraeff, G. (1992). Toxic composition of resting cyst of *Alexandrium tamarense* (Dinophyceae). *Toxicon, 30*, 1539–1544.

Ottway, B., Parker, M., & McGrath, D. (1979). Observations on a bloom of Gyrodiniumaureolum Hulbert on the south coast of Ireland, summer 1976, associated with mortalities of littoral and sub-littoral organisms. *Fish Sem* Ser., B (18).

Paerl, H. W., Bales, J. D., Ausley, L. W., Buzzelli, C. P., Crowder, L. B., Eby, L. A., et al. (2001a). Ecosystem impacts of three sequential hurricanes (Dennis, Floyd, and Irene) on the United States' largest lagoonal estuary, Pamlico Sound, NC. *Proceedings of the National Academy of Sciences, 98*(10), 5655–5660.

Paerl, H. W., Fulton, R. S., Moisander, P. H., & Dyble, J. (2001b). Harmful freshwater algal blooms, with an emphasis on cyanobacteria. *ScientificWorldJournal, 1*, 76–113.

Parab, S. G., Matondkar, S. P., Gomes, H. D. R., & Goes, J. I. (2006). Monsoon driven changes in phytoplankton populations in the eastern Arabian Sea as revealed by microscopy and HPLC pigment analysis. *Continental Shelf Research, 26*(20), 2538–2558.

Park, J., Kim, H., & Lee, S. (1989). Studies on red tide phenomena in Korean coastal waters. In T. Okaichi, D. M. Anderson, & T. Nemoto (Eds.), *Red tides: Biology, environmental science, and toxicology* (pp. 37–40). New York: Elsevier.

Park, T. G., Lim, W. A., Park, Y. T., Lee, C. K., & Jeong, H. J. (2013). Economic impact, management and mitigation of red tides in Korea. *Harmful Algae, 30*(Suppl 1), S131–S143.

Parrish, C. C. (1987). Time series of particulate and dissolved lipid classes during spring phytoplankton blooms in Bedford Basin, a marine inlet. *Marine Ecology Progress Series, 35*(129–139), 10.

Parsons, M. L., & Dortch, Q. (2002). Sedimentological evidence of an increase in Pseudo-nitzschia (Bacillariophyceae)abundance in response to coastal eutrophication. *Limnology and Oceanography, 47*(2), 551–558. 1939-5590. https://doi.org/10.4319/lo.2002.47.2.0551.

Parsons, M. L., Scholin, C. A., Miller, P. E., Doucette, G. J., Powell, C. L., Fryxell, G. A., Dortch, Q., & Soniat, T. M. (1999). Pseudo-nitzschia species (Bacillariophyceae) in Louisiana coastal waters: Molecular probe field trials, genetic variability, and domoic acid analyses. *Journal of Phycology, 35*, 1368–1378.

Perl, T. M., Bedard, L., Kosatsky, T., Hockin, J. C., Todd, E. C. D., McNutt, L. A., & Remis, R. S. (1990a). *Amnesic shellfish poisoning: A new clinical syndrome due to domoic acid.* In: I. Hynie & E. C. D. Todd (Eds.), Proceedings of a symposium, domoic acid toxicity. Canada disease weekly report, Ottawa, Ontario, pp. 7–8.

Perl, T. M., Bédard, L., Kosatsky, T., Hockin, J. C., Todd, E. C. D., & Remis, R. S. (1990b). An outbreak of toxic encephalopathy caused by eating mussels contaminated with domoic acid. *New England Journal of Medicine, 322*(25), 1775–1780.

Persson, A., Smith, B. C., Alix, J. H., Senft-Batoh, C., & Wikfors, G. H. (2012). Toxin content differs between life stages of *Alexandrium fundyense* (Dinophyceae). *Harmful Algae, 19*, 101–107.

Pierce, R. H., & Henry, M. S. (2008). Harmful algal toxins of the Florida red tide (*Karenia brevis*): Natural chemical stressors in South Florida coastal ecosystems. *Ecotoxicology, 17*(7), 623–631.

Pingree, R. D., Pugh, P. R., Holligan, P. I., & Forster, G. R. (1975). Summer phytoplankton blooms and red tides along tidal fronts in the approaches to the English Channel. *Nature, 258*(5537), 672–677.

Planque, B., & Taylor, A. H. (1998). Long-term changes in zooplankton and the climate of the North Atlantic. *ICES Journal of Marine Science, 55*(4), 644–654.

Poli, M. A., Musser, S. M., Dickey, R. W., Eilers, P. P., & Hall, S. (2000). Neurotoxic shellfish poisoning and brevetoxin metabolites: A case study from Florida. *Toxicon, 38*(7), 981–993.

Porumb, F. (1992). On the development of *Noctiluca scintillans* under eutrophication of Romanian Black sea waters. *The Science of Total Environment, 126*(Suppl), 907–920.

Poulin, R. X., Poulson-Ellestad, K. L., Roy, J. S., & Kubanek, J. (2018). Variable allelopathy among phytoplankton reflected in red tide metabolome. *Harmful Algae, 71*, 50–56.

Pratchett, M. S., Munday, P., Wilson, S. K., Graham, N. A., Cinner, J. E., Bellwood, D. R., et al. (2008). Effects of climate-induced coral bleaching on coral-reef fishes. *Ecological and economic consequences. Oceanography and Marine Biology: An Annual Review, 46*, 251–296.

Rabalais, N. N., Turner, R. E., Díaz, R. J., & Justić, D. (2009). Global change and eutrophication of coastal waters. *ICES Journal of Marine Science, 66*(7), 1528–1537.

Rathaille, A. N., & Raine, R. (2011). Seasonality in the excystment of Alexandrium minutum and Alexandrium tamarense in Irish coastal waters. *Harmful Algae, 10*(6), 629–635.

Reboreda, A., Lago, J., Chapela, M. J., Vieites, J. M., Botana, L. M., Alfonso, A., & Cabado, A. G. (2010). Decrease of marine toxin content in bivalves by industrial processes. *Toxicon, 55*, 235–243.

Reddy, P. R. C., Salvekar, P. S., & Nayak, S. (2008). Super cyclone induces a mesoscale phytoplankton bloom in the bay of Bengal. *IEEE Geoscience and Remote Sensing Letters, 5*(4), 588–592.

Reich, A., Lazensky, R., Faris, J., Fleming, L. E., Kirkpatrick, B., Watkins, S., Ullmann, S., Kohler, K., & Hoagland, P. (2015). Assessing the impact of shellfish harvesting area closures on neurotoxic shellfish poisoning (NSP) incidence during red tide (*Karenia brevis*) blooms. *Harmful Algae, 43*, 13–19.

Richlen, M. L., Morton, S. L., Jamali, E. A., Rajan, A., & Anderson, D. M. (2010). The catastrophic 2008–2009 red tide in the Arabian gulf region, with observations on the identification and phylogeny of the fish-killing dinoflagellate Cochlodinium polykrikoides. *Harmful Algae, 9*(2), 163–172.

Robin, R., Kanuri, V. V., Muduli, P. R., Mishra, R. K., Jaikumar, M., Karthikeyan, P., Suresh Kumar, C., & Saravana Kumar, C. (2013). Dinoflagellate bloom of *Karenia mikimotoi* along the Southeast Arabian Sea, bordering Western India. *Journal of Ecosystems, 2013*, 463720.

Rockström, J., Steffen, W., Noone, K., Persson, A., Chapin, F. S., III, Lambin, E. F., Lenton, T. M., Scheffer, M., Folke, C., Schellnhuber, H. J., Nykvist, B., de Wit, C. A., Hughes, T., van der Leeuw, S., Rodhe, H., Sörlin, S., Snyder, P. K., Costanza, R., Svedin, U., Falkenmark, M., Karlberg, L., Corell, R. W., Fabry, V. J., Hansen, J., Walker, B., Liverman, D., Richardson, K., Crutzen, P., & Foley, J. A. (2009). A safe operating space for humanity. *Nature, 461*, 472–475.

Rodrigue, D. C., Etzel, R. A., Hall, S., de Porras, E., Velasquez, O. H., Tauxe, R. V., Kilbourne, E. M., & Blake, P. A. (1990). Lethal paralytic shellfish poisoning in Guatemala. *The American Journal of Tropical Medicine Hygiene, 42*(3), 267–271.

Rodríguez, F., Fraga, S., Ramilo, I., Rial, P., Isabel, R., Figueroa, R., Riobó, P., & Bravo, I. (2017). Canary Islands (NE Atlantic) as a biodiversity 'hotspot' of Gambierdiscus: Implications for future trends of ciguatera in the area. *Harmful Algae, 67*, 131–143.

Rongo, T., & van Woesik, R. (2012). Socioeconomic consequences of ciguatera poisoning in Rarotonga, southern Cook Islands. *Harmful Algae, 20*, 92–100.

Roué, M., Darius, H. T., Viallon, J., Ung, A., Gatti, C., Harwood, D. T., & Chinain, M. (2018). Application of solid phase adsorption toxin tracking (SPATT) devices for the field detection of *Gambierdiscus* toxins. *Harmful Algae, 71*, 40–49.

Salas, R., Tillmann, U., John, U., Kilcoyne, J., Burson, A., Cantwell, C, Hess, P., Jauffrais, T., & Silke, J. (2011). The role of *Azadinium spinosum* (Dinophyceae) in the production of azaspiracid shellfish poisoning in mussels. *Harmful Algae, 10*(6), 774–783. doi:https://doi.org/10.1016/j.hal.2011.06.010. ISSN: 1568-9883.

Sar, E. A., Sunesen, I., & Jahn, R. (2010). Coscinodiscus perforatus Revisited and compared with *Coscinodiscus radiatus* (Bacillariophyceae). *Phycologia, 49*(6), 514–524.

Satake, M., Shoji, M., Oshima, Y., Naoki, H., Fujita, T., & Yasumoto, T. (2002). Gymnocin- A, a cytotoxic polyether from the notorious red tide dinoflagellate, *Gymnodinium mikimotoi*. *Tetrahedron Letters, 43*(33), 5829–5832.

Satake, M., Tanaka, Y., Ishikura, Y., Oshima, Y., Naoki, H., & Yasumoto, T. (2005). Gymnocin-B with the largest contiguous polyether rings from the red tide dinoflagellate, *Karenia* (formerly *Gymnodinium) mikimotoi*. *Tetrahedron Letters, 46*(20), 3537–3540.

Schnetzer, A., Lampe, R. H., Benitez-Nelson, C. R., Marchetti, A., Osburn, C. L., & Tatters, A. O. (2017). Marine snow formation by the toxin-producing diatom, *Pseudo-nitzschia australis*. *Harmful Algae, 61*, 23–30.

Scholin, C., Doucette, G., Jensen, S., Roman, B., Pargett, D., Marin, R., III, Preston, C., Jones, W., Feldman, J., Everlove, C., Harris, A., Alvarado, N., Massion, E., Birch, J., Greenfield, D., Vrijenhoek, R., Mikulski, C., & Jones, K. (2009). Remote detection of marine microbes, small invertebrates, harmful algae, and biotoxins using the Environmental Sample Processor (ESP). *Oceanography, 2*, 158–167.

Schütt, F. (1895). Peridineen der Plankton-Expedition. Ergebn. Plankton–Expedition der Humboldt-Stiftung 4 (M, a, A) 1–170, 127 pls.

Sekula-Wood, E., Benitez-Nelson, C., Morton, S., Anderson, C., Burrell, C., & Thunell, R. (2011). Pseudo-nitzschia and domoic acid fluxes in Santa Barbara Basin (CA) from 1993 to 2008. *Harmful Algae, 10*(6), 567–575.

Sellner, K. G. (1997). Physiology, ecology and toxic properties of marine cyanobacteria blooms. *Limnology and Oceanography, 42*, 1089–1104.

Sengco, M. R., & Anderson, D. M. (2004). Controlling harmful algal blooms through clay flocculation. *Journal of Eukaryotic Microbiology, 51*(2), 169–172.
Sengco, M. R., Li, A. S., Tugend, K., Kulis, D., & Anderson, D. M. (2001). Removal of red- and brown-tide cells using clay flocculation. I. Laboratory culture experiments with *Gymnodinium breve* and *Aureococcus anophagefferens*. *Marine Ecology Progress Series, 210*, 41–53.
Shen, L., Xu, H., & Guo, X. (2012). Satellite remote sensing of Harmful Algal Blooms (HABs) and a potential synthesized framework. *Sensors (Basel, Switzerland), 12*(6), 7778–7803.
Sheng, J., Malkiel, E., Katz, J., Adolf, J. E., & Place, A. R. (2010). A dinoflagellate exploits toxins to immobilize prey prior to ingestion. *Proceedings of the National Academy of Sciences of the United States of America, 107*(5), 2082–2087.
Shuler, A. J., Paternoster, J., Brim, M., Nowocin, K., Tisdale, T., Neller, K., Cahill, J. A., Leighfield, T. A., SpencerFire, S., Wang, Z., & Morton, S. (2012). Spatial and temporal trends of the toxic diatom, *Pseudo-nitzschia* in the southeastern Atlantic United States. *Harmful Algae, 17*, 6–13.
Shumway, S. E. (1990). A review of the effects of algal blooms on shellfish and aquaculture. *Journal of the World Aquaculture Society, 21*, 65–104.
Shutler, J. D., Davidson, K., Miller, P. I., Swan, S. C., Grant, M. G., & Bresnan, E. (2012). An adaptive approach to detect high-biomass algal blooms from EO chlorophyll-a data in support of harmful algal bloom monitoring. *Remote Sensing Letters, 3*(2), 101–110.
Silke, J., O'Beirn, F., & Cronin, M. (2005). *Kareniamikimotoi: An exceptional Dino- flagellate bloom in western Irish waters, summer 2005*. Galway: Marine Institute.
Silva, E. S. (1967). *Cochlodinium heterolobactum* n. Sp.: Structure and some cytophysiological aspects. *The Journal of Protozoology, 14*, 745–754.
Sim, J., & Wilson, N. (1997). Surveillance of marine biotoxins, 1993–1996. *Public Health Reports, 4*, 9–16.
Smayda, T. J. (2006). *Harmful algal bloom communities in Scottish coastal waters: Relationship to fish farming and regional comparisons – A review*. Natural Scotland Scottish Executive.
Soto, I. M., Cambazoglu, M. K., Boyette, A. D., Broussard, K., Sheehan, D., Howden, S. D., & Arnone, R. A. (2018). Advection of *Karenia brevis* blooms from the Florida panhandle towards Mississippi coastal waters. *Harmful Algae, 72*, 46–64.
Steidinger, K. A. (2009). Historical perspective on Karenia brevis red tide research in the Gulf of Mexico. *Harmful Algae, 8*, 549–561.
Steidinger, K. A., & Tangen, K. (1996). Dinoflagellates. In C. R. Tomas (Ed.), *Identifying marine diatoms and dinoflagellates* (pp. 387–598). New York: Academic.
Sun, X. X., Han, K. N., Choi, J. K., & Kim, E. K. (2004a). Screening of surfactants for harmful algal blooms mitigation. *Marine Pollution Bulletin, 48*(9), 937–945.
Sun, X. X., Lee, Y. J., Choi, J. K., & Kim, E. K. (2004b). Synergistic effect of sophorolipid and loess combination in harmful algal blooms mitigation. *Marine Pollution Bulletin, 48*(9), 863–872.
Sunda, W., Graneli, E., & Gobler, C. (2006). Positive feedback and the development and the persistence of ecosystem disruptive algal blooms. *Journal of Phycology, 42*, 963–974.
Sweeney, B. M. (1978). Ultrastructure of *Noctiluca miliaris* (Pyrrophyta) with green symbionts. *Journal of Phycology, 14*, 116–120.
Taleb, H., Vale, P., Amanhir, R., Benhadouch, A., Sagou, R., & Chafik, A. (2006). First detection of azaspiraicds in mussels in north West Africa. *Journal of Shellfish Research, 25*, 1067–1070.
Tang, Y. Z., & Gobler, C. J. (2012). The toxic dinoflagellate *Cochlodinium polykrikoides* (Dinophyceae) produces resting cysts. *Harmful Algae, 20*, 71–80.
Tang, D., Di, B., Wei, G., Ni, I. H., & Wang, S. (2006). Spatial, seasonal and species variations of harmful algal blooms in the south Yellow Sea and East China Sea. *Hydrobiologia, 568*, 245–253.
Tangen, K. (1977). Blooms of *Gyrodinium aureolum* (Dinophyeae) in North European waters, accompanied by mortality in marine organisms. *Sarsia, 63*(2), 123–133.
Taylor, F. J. R. (1976). *Dinoflagellates from the International Indian Ocean Expedition*. A report on material collected by the R.V. "Anton Bruun" 1963–1964. Bibl. Bot. 132. Schweizerbart'sche Verlagsbuchhandlung, Stuttgart. 234 pp., pls. 46.

Taylor, F. J. R. (1993). The species problem and its impact on harmful phytoplankton studies, with emphasis on dinoflagellate morphology. In T. J. Smayda & Y. Shimizu (Eds.), *Toxic phytoplankton blooms in the sea* (pp. 81–86). Amsterdam: Elsevier.

Taylor F. J. R., Fukuyo Y., & Larzen J., (1995). Taxonomy of harmful Dinoflagellates. In G. M. Hallegraeff, D. M. Anderson & A. D. Cembella (Eds.), *Manual on harmful marine microalgae* (Monographs on oceanographic methodology, pp. 283–317). Paris: UNESCO.

Teitelbaum, J. (1990). *Clinical presentation of acute intoxication by domoic acid: Case observations*. In I. Hynie & E. C. D Todd (Ed.), Proceedings of a symposium, Domoic acid toxicity. Canada disease weekly report, Ottawa, ON, pp. 5–6.

Teitelbaum, J. S., Zatorre, R. J., Carpenter, S., Genderon, D., Evans, A. C., Gjedde, A., & Cashman, N. R. (1990). Neurological sequelae of domoic acid intoxication due to the ingestion of contaminated mussels. *The New England Journal of Medicine, 322*, 1781–1787.

Tiling, K., & Proffitt, C. E. (2017). Effects of Lyngbya majuscula blooms on the seagrass Halodule wrightii and resident invertebrates. *Harmful Algae, 62*, 104–112.

Tillmann, U., Elbrächter, M., Krock, B., John, U., & Cembella, A. (2009). *Azadinium spinosum-gen.* et sp. nov. (Dinophyceae) identified as a primary producer of azaspiracid toxins. *European Journal of Phycology, 44*(1), 63–79. 0967-0262. https://doi.org/10.1080/09670260802578534.

Tillmann, U., Jaen, D., Fernandez, L., Gottschling, M., Witt, M., Blanco, J., & Krock, B. (2017). *Amphidoma languida* (Amphidomatacea, Dinophyceae) with a novel azaspiracid toxin profile identified as the cause of molluscan contamination at the Atlantic coast of southern Spain. *Harmful Algae, 62*, 113–126.

Tomas, C. R., & Smayda, T. J. (2008). Red tide blooms of *Cochlodinium polykrikoides* in a coastal cove. *Harmful Algae, 7*(3), 308–317.

Torgersen, T., Bremnes, N. B., Rundberget, T., & Aune, T. (2008). Structural confirmation and occurrence of azaspiracids in Scandinavian brown crabs (*Cancer pagurus*). *Toxicon, 51*, 93–101.

Touzet, N., & Raine, R. (2007). Discrimination of *Alexandrium andersoni* and *A. minutum* (Dinophyceae) using LSU rRNA-targeted oligonucleotide probes and fluorescent whole-cell hybridization. *Phycologia, 46*(2), 168–177.

Trainer, V. L., Moore, L., Bill, B. D., Adams, N. G., Harrington, N., Borchert, J., et al. (2013). Diarrhetic shellfish toxins and other lipophilic toxins of human health concern in Washington state. *Marine Drugs, 11*(6), 1815–1835.

Turkoglu, M. (2010). Temporal variations of surface phytoplankton, nutrients and chlorophyll-a in the Dardanelles (Turkish Straits System): A coastal station sample in weekly time intervals. *Turkish Journal of Biology, 34*(3), 319–333.

Turkoglu, M., & Buyukates, Y. (2005). Short time variations in density and bio-volume of *Noctiluca scintillans* (Dinophyceae) in Dardanelles, XIII. Natnl. Fish. Symp., 01–04 Semptember 2005, C¸ anakkale, Turkey, Abstr. Book (Abstracts), 59 (in Turkish).

Turkoglu, M., & Erdogan, Y. (2010). Diurnal variations of summer phytoplankton and interactions with some physicochemical characteristics under eutrophication of surface water in the Dardanelles (C¸ anakkale Strait, Turkey). *Turkish Journal of Biology, 34*(2), 211–225.

Turkoglu, M., & Koray, T. (2002). Phytoplankton species succession and nutrients in southern Black Sea (Bay of Sinop). *Turkish Journal of Botany, 26*, 235–252.

Turkoglu, M., & Koray, T. (2004). Algal blooms in surface waters of the Sinop Bay in the Black Sea, Turkey. *Pakistan Journal of Biological Sciences, 7*(9), 1577–1585. https://doi.org/10.3923/pjbs.2004.1577.1585.

Turkoglu, M., Unsal, M., Işmen, A., Mavili, S., Sever, T. M., Yenici, E., Kaya, S. & Çoker, T. (2004). *Dinamic of lower and high food chain of the Dardanelles and Saros Bay (North Aegean Sea)*. TUBITAK-ÇAYDAG Technical Final Report (p. 313), No: 101Y081, Çanakkale, Turkey.

Turrell, E., Bresnan, E., Collins, C., Brown, L., Graham, J., & Grieve, M. (2008). Detection of *Pseudo-nitzschia* (Bacillariophyceae) species and amnesic shellfish toxins in Scottish coastal waters using oligonucleotide probes and the Jellet rapid test™. *Harmful Algae, 7*(4), 443–458.

Twiner, M. J., Rehmann, N., Hess, P., & Doucette, G. J. (2008). Azaspiracid shellfish poisoning: A review on the chemistry, ecology and toxicology with an emphasis on human health impact. *Marine Drugs, 6*, 39–72.

Ulloa, M. J., Álvarez-Torres, P., Horak-Romo, K. P., & Ortega-Izaguirre, R. (2017). Harmful algal blooms and eutrophication along the Mexican coast of the Gulf of Mexico large marine ecosystem. *Environmental Development, 22*, 120–128.

Unsal, M., Turkoglu, M., & Yenici, E. (2003). *Biological and physicochemical researches in the Dardanelles*. TUBITAK-YDABAG Technicial Final Report, 101Y075, Canakkale, 92 pp. (in Turkish).

Urquhart, E. A., Schaeffer, B. A., Stumpf, R. P., Loftin, K. A., & Jeremy Werdell, P. (2017). A method for examining temporal changes in cyanobacterial harmful algal bloom spatial extent using satellite remote sensing. *Harmful Algae, 67*, 144–152.

Valdiglesias, V., Laffon, B., Pásaro, E., & Méndez, J. (2011). Okadaic acid induces morphological changes, apoptosis and cell cycle alterations in different human cell types. *Jounal of Environmental Monitoring, 13*(6), 1831–1840.

Vale, P. (2004). Is there a risk of human poisoning by azaspiracids from shellfish harvested at the Portuguese coast? *Toxicon, 44*, 943–947.

Vale, P. J., Botelho, M. J., Rodrifues, S. M., Gomes, S. S., Sampayo, d. M., & A, M. (2008). Two decades of marine biotoxin monitoring in bivalves from Portugal (1986–2006): A review of exposure assessment. *Harmful Algae, 7*, 11–25.

Van der Lingen, C. D., Hutchings, L., Lamont, T., & Pitcher, G. C. (2016). Climate change, dinoflagellate blooms and sardine in the southern Benguela current large marine ecosystem. *Environmental Development, 17*, 230–243.

van Deventer, M., Atwood, K., Vargo, G. A., Flewelling, L. J., Landsberg, J. H., Naar, J. P., & Stanek, D. (2012). Karenia brevis red tides and brevetoxin-contaminated fish: A high risk factor for Florida's scavenging shorebirds? *Botanica Marina, 55*, 31–37.

Van Egmond, H. P., Aune, T., Lassus, P., Speijers, G. J. A., & Paralytic, W. M. (1993). Diarrhoeic shellfish poisons: Occurrence in Europe, toxicity, analysis, and regulation. *Journal of Natural Toxins, 2*(1), 41–83.

Vanhoutte-Brunier, A., Fernand, L., Ménesguen, A., Lyons, S., Gohin, F., & Cugier, P. (2008). Modelling the Karenia mikimotoi bloom that occurred in the western English Channel during summer 2003. *Ecological Modelling, 210*(4), 351–376.

Vogelbein, W. K., Lovko, V. J., Shields, J. D., Reece, K. S., Mason, P. L., Haas, L. W., & Walker, C. C. (2002). Pfiesteria shumwayae kills fish by micropredation not exotoxin secretion. *Nature, 418*(6901), 967–970.

Walker, N. D., Leben, R. R., & Balasubramanian, S. (2005). Hurricane-forced upwelling and chlorophyll a enhancement within cold-core cyclones in the Gulf of Mexico. *Geophysical Research Letters, 32*(18), L18610.

Wang, Y., Pan, J., Cai, J., & Zhang, D. (2012). Floating assembly of diatom *Coscinodiscus* sp. microshells. *Biochemical and Biophysical Research Communications, 420*, 1–5.

Wang, Y., Yang, L., Kong, L., Liuc, E., Wang, L., & Zhu, J. (2015). Spatial distribution, ecological risk assessment and source identification for heavy metals in surface sediments from Dongping Lake, Shandong, East China. *Catena, 125*, 200–205.

Watkins, S. M., Reich, A., Fleming, L. E., & Hammond, R. (2008). Neurotoxin shell poisoning. *Marine Drugs, 6*, 431–455.

Wekell, J. C., Gauglitz, E. J., Bamett, H. J., Hatfield, C. L., Simons, D., & Ayres, D. (1994). Occurrence of domoic acid in Washington state razor clams (*Siliqua patula*) during 1991-1993. *Natural Toxins, 2*(4), 197–205.

Whyte, J. N. C., Haigh, N., Ginther, N. G., & Keddy, L. J. (2001). First record of blooms of *Cochlodinium* sp. (Gymnodiniales, Dinophyceae) causing mortality to aquacultured salmon on the west coast of Canada. *Phycologia, 40*, 298–304.

Whyte, C., Swan, S., & Davidson, K. (2014). Changing wind patterns linked to unusually high Dinophysis blooms around the Shetland Islands, Scotland. *Harmful Algae, 39*, 365–373.

Wong, P. S. (1989). The occurrence and distribution of red tides in Hong Kong – Applications in red tide management. In T. Okaichi, D. M. Anderson, & T. Nemoto (Eds.), *Red tides: Biology, environmental science and toxicology* (pp. 125–128). New York: Elsevier Science Publishers.

Yasumoto, T., Oshima, Y., & Yamaguchi, M. (1978). Occurrence of a new type of toxic shellfish poisoning in the Tohoku district. *Bulletin of the Japanese Society of Scientific Fisheries, 44*, 1249–1255.

Yin, K. (2003). Influence of monsoons and oceanographic processes on red tides in Hong Kong waters. *Marine Ecology Progress Series, 262*, 27–41.

Yoshimatsu, S. (1984). Sexual reproduction of *Protogonyaulax catenella* in culture. II. Mating type. *Bulletin of the Plankton Society of Japan, 31*, 107–111.

Yu, Z. M., & Zou, J. Z. (1994). A more effective clay for remove red tide organisms. *Jounal of Natural Disasters, 3*(2), 105–109. (in Chinese).

Yu, Z. M., Zou, J. Z., & Ma, X. N. (1994a). Application of clays to removal of red tide organisms I. Coagulation of red tide organisms with clays. *Chinese Journal of Oceanology and Limnology, 12*(3), 193–200.

Yu, Z. M., Zou, J. Z., & Ma, X. N. (1994b). Application of clays to removal of red tide organisms II. Coagulation of different species of red tide organisms with montmorillonite and effect of clay pretreatment. *Chinese Journal of Oceanology and Limnology, 12*(4), 316–324.

Yu, Z. M., Zou, J. Z., & Ma, X. N. (1994c). A new method to improve the capability of clays for removing red tide organisms. *Oceanologia et Limnologia Sinica, 25*(2), 226–232.

Yu, Z. M., Ma, X. N., & Xie, Y. (1995). Study of main nutrients adsorption on clays in seawater. *Oceanology and Limnology Sinica, 26*(2), 208–214 (in Chinese).

Yu, Z. M., Sun, X. X., Song, X. X., & Zhang, B. (1999). Clay surface modification and its coagulation of red tide organisms. *Chinese Science Bulletin, 44*(7), 617–620 (in Chinese).

Yuan, J., Mi, T., Zhen, Y., & Zhigang, Y. (2012). Development of a rapid detection and quantification method of *Karenia mikimotoi* by real-time quantitative PCR. *Harmful Algae, 17*, 83–91.

Zhang, Z., Yang, X., Li, H., Li, W., Yan, H., & Shi, F. (2017). Application of a novel hybrid method for spatiotemporal data imputation: A case study of the Minqin County groundwater level. *Journal of Hydrology, 553*, 384–397.

Zhang, Y., Yu, Z., Song, X., Yuan, Y., & Cao, X. (2018). Effects of modified clay used for the control of harmful algal blooms on Alexandrium pacificum cysts. *Harmful Algae, 72*, 36–45.

Zheng, G. M., & Tang, D. L. (2007). Offshore and nearshore chlorophyll increases induced by typhoon winds and subsequent terrestrial rainwater runoff. *Marine Ecology Progress Series, 333*, 61–74.

Zhu, M., Paerl, H. W., Zhu, G., Wu, T., Li, W., Shi, K., & Caruso, A. M. (2014). The role of tropical cyclones in stimulating cyanobacterial (*Microcystis* spp.) blooms in hypertrophic Lake Taihu, China. *Harmful Algae, 39*, 310–321.

Zingmark, R. G. (1970). Sexual reproduction in the dinoflagellate *Noctiluca miliaris* Suriray. *Journal of Phycology, 6*, 122–126.

Printed in the United States
By Bookmasters